ADVANCED TOPICS IN JUST-IN-TIME MANAGEMENT

ADVANCED TOPICS IN JUST-IN-TIME MANAGEMENT

MARC J. SCHNIEDERJANS
JOHN R. OLSON

QUORUM BOOKS
Westport, Connecticut • London

Library of Congress Cataloging-in-Publication Data

Schniederjans, Marc J.
 Advanced topics in just-in-time management / Marc J.
 Schniederjans, John R. Olson.
  p. cm.
 Includes bibliographical references and index.
 ISBN 1–56720–155–5 (alk. paper)
 1. Just-in-time systems. I. Olson, John R., 1938– .
 II. Title.
 TS157.4.S38 1999
 658.5′6—dc21   99–21590

British Library Cataloguing in Publication Data is available.

Library of Congress Catalog Card Number: 99–21590
ISBN: 1–56720–155–5

First published in 1999

Quorum Books, 88 Post Road West, Westport, CT 06881
An imprint of Greenwood Publishing Group, Inc.
www.quorumbooks.com

Printed in the United States of America

The paper used in this book complies with the
Permanent Paper Standard issued by the National
Information Standards Organization (Z39.48–1984).

10 9 8 7 6 5 4 3 2 1

To Adolph Olson and Donna Olson,
truly superior parents

CONTENTS

TABLES AND FIGURES

TABLES

FIGURES

PREFACE

Have you been using Just-In-Time (JIT) methods and feel the need to go to the next level up in JIT thinking? Feel your JIT approach is a bit rusty and you need to find out what advances are being made in this strategy for world-class manufacturing? Are you curious about what successful JIT practitioners have been doing with JIT in the last few years? Wonder what new JIT idea academicians are researching and theorizing that may make the JIT revolution new all over again? Regardless of whether you are a vice president of operations or a shop floor supervisor, if you are using JIT or have any interest in the subject, this book is for you. The focus of this book is to help make you aware of newest, most advanced issues, concepts, procedures, methodologies, and practices on the implementation and use of JIT. This book will be of use to industrial engineers, process engineers, operations managers, inventory managers, production schedulers, material handling managers, and logistic planners. Operations management faculty, trainers, and graduate students will also find new directions for JIT research in production and service organizations.

This book assumes that the reader is an educated operations practitioner or student with at least one year of exposure to the terminology commonly found in JIT. Terms not defined in this book can be found in either the 1998, 9th edition *American Production and Inventory Control Society* (APICS) *Dictionary* or in the 1998 *Encyclopedia of Production and Manufacturing Management*. Throughout this book, important terms are italicized and usually be followed by a definition. The location of the initial definitions can found using the index at the end of the book.

The contents of this book are organized into three parts. In Part I, "Introduction," a single chapter is presented that defines JIT ideas used in this book and explains how the chapters are structured. Part II, "Just-In-Time Quantitative Tools" contains three chapters. These chapters are structured to present methodology and discussion on the very latest advances in quantitative methods that have been reported in the literature. Part III, "Selected Just-In-Time Topics," contains four chapters that focus on important topics that have made a significant impact in the successfulness of JIT operations.

To support the material presented in the chapters, two appendices are included at the end of book. Appendix I, "Sources of Additional JIT Book Information," provides a bibliography of JIT books that have appeared in the literature during the last 20 years. Appendix II, "Sources of Additional JIT Journal Article Information," provides a bibliography of JIT articles that where published during the last 15 years. This appendix divides the articles into groupings that are related to the topics in the book and other important topical areas. These divisional groupings also permit easier access to the specific information that JIT practitioners might need for quick reference.

While many people have had a hand in the preparation of this book, its accuracy and completeness are our responsibility.

PART I

INTRODUCTION

1

INTRODUCTION TO JUST-IN-TIME MANAGEMENT

The purpose of this chapter is to establish a baseline of what Just-In-Time (JIT) management is about. This chapter also identifies the basic theme of the book and describes its organization to help readers understand its content. Selected JIT terminology used throughout the book is also introduced in this chapter.

WHAT IS JUST-IN-TIME MANAGEMENT?

When the term *Just-In-Time* or *JIT* is used in management literature, it can mean different things to different people (see Vokurka and Davis, 1996). JIT is considered an "operational" (i.e., shop-floor) management approach to minimizing waste in operations (see Keane and King, 1991). JIT is also considered a "tactical" management concept to improving efficiency and product quality (see Sakakibara, Flynn, and Morris, 1997). To still others, JIT is "strategic" management approach to achieve world-class manufacturing (see Grieco, 1990; Schonberger, 1986, 1996). In practice, JIT is more than a operational, tactical, or strategic approach to running an operation; it is a philosophy that can be applied to virtually every organization (manufacturing or service) and to every aspect of business life (see Vokurka and Davis, 1996).

For the purpose of this book, JIT management will be defined as it has come to be known as a set of need-to-know principles for successfully operating a competitive business (see Tables 1.1 through 1.6). Historically these principles migrated from Japan and chiefly started (see Keller and Kazazi, 1993) in the United States during the late 1970s and early 1980s in the manufacturing area of inventory management (see Table 1.1 and Hall, 1986). The principles quickly expanded to include the areas of production management (see Table 1.2 and Merli, 1990; Rutherford, 1981) and quality management (see Table 1.3 and Bare, 1991). As they were successfully applied internally within the organization, the need to incorporate external partners, like suppliers and vendors motivated still more JIT principles to live by (see Table 1.4 and Ansari and Modarress, 1990; Harding, 1990; Schorr, 1998). During the mid-to-late 1980s a reexamination of JIT motivated a greater emphasis on the human resource component in JIT (see Table 1.5 and Adair-Heeley, 1991) and the physical work environment in which the human resources operated (see Table 1.6 and Black, 1991; Hirano, 1988). During this same time, JIT principles were also extended into service operations and administrative functions (see Schniederjans, 1993, pp. 234 -264).

TABLE 1.1. JIT Inventory Management Principles

- Seek zero inventory.
- Seek reliable suppliers.
- Seek reduced buffer inventory.
- Seek reduced lot sizes and increased frequency of orders.
- Seek improved inventory handling.
- Seek to continuously identify and correct all inventory problems.

TABLE 1.2. JIT Production Management Principles

- Seek uniform daily production scheduling.
- Seek production scheduling flexibility.
- Seek a mixed model scheduling system.
- Seek a synchronized pull system and eliminate waste.
- Seek improved flexibility in providing product changeovers.
- Seek improved communication and visualization.
- Seek reduced production lot sizes.
- Seek reduced production setup costs.
- Allow workers to determine production flow.
- Seek unitary production.
- Schedule work at less than full capacity.
- Increase standardization of product processing.
- Seek to continuously identify and correct all production management problems.

TABLE 1.3. JIT Quality Management Principles

- Seek long-term commitment to quality control efforts.
- Seek high level of visibility management on quality.
- Use fail safe methods to help ensure quality conformity.
- Maintain process control and strict product quality compliance.
- Utilize statistical quality control methods to monitor and motivate product quality.
- Seek to empower workers by sharing authority in the control of product quality.
- Require self-correction of worker-generated defects.
- Require workers to perform routine maintenance and housecleaning duties.
- Seek to make quality everybody's responsibility.
- Maintain 100 percent quality inspection of products through *work-in-process* (WIP) efforts.
- Seek to continuously identify and correct all quality related problems.

TABLE 1.4. JIT Supplier Relation Principles

- Seek certification in quality of items purchased.
- Seek improved quality.
- Seek reduced costs.
- Seek timely communications and responsiveness.
- Seek smaller lots with more frequent delivery.
- Seek single-source suppliers.
- Seek long-term relationships with suppliers.
- Seek ordering flexibility.
- Seek to reduce inventory delivery lead time.
- Seek to continuously identify and correct all supplier relation problems.

Some of the JIT principles in Tables 1.1 through 1.6 have been stated in different ways, at different times, and in different countries. For example, *kaizen* is a Japanese term applied in Japan to finding and eliminating waste in all aspects of manufacturing (see Vokurka and Davis, 1996). Yet JIT is also a waste removing approach, so JIT and kaizen are really subset topics. It can also be seen in the tables that JIT involves a principle of *continuous improvement* (CI). Yet, CI is more commonly associated with *total quality management* (TQM). So, JIT principles can be viewed in part as a subset of TQM. Some familiarity with CI and TQM will help readers to understand this book's content.

TABLE 1.5. JIT Human Resource Principles

- Provide continuous and extensive training.
- Seek long-term or lifetime commitment to employees.
- Seek a highly flexible workforce.
- Maintain a substantial part-time workforce.
- Employee team approach to production cells and problem solving.
- Seek to establish a family atmosphere to build trust in employees.
- Utilize peer pressure to motivate employee performance.
- Seek to build pride in workmanship and mutual respect of employees.
- Establish compensation plans that reward individual and team efforts.
- Seek to continuously identify and correct all human resource problems.

TABLE 1.6. JIT Facility Design Principles

- Use automation (i.e., robots, computer integrated manufacturing, etc.) where practical.
- Seek a focused factory.
- Identify and eliminate production bottlenecks .
- Use group technology cells (i.e., U-line or C-cells) in production layouts.
- Seek to maximize flow through layout .
- Reduce distance between all production activities.
- Reallocate reduced physical space under JIT to other productive activities.
- Seek to minimize material handling by using reusable containers and unity packaging.
- Seek to minimize material flow congestion by designing replenishment systems closer to the point of use.
- Seek continuous redesign efforts to improve facility layout and facility structure.

As sited, the JIT principles presented in Tables 1.1 through 1.6 have been collected from a variety of past and current JIT books and articles. Collectively, they represent the type of basic information in which a JIT manager should be fairly familiar. (For a summary review of each these principles, see Cheng and Podolsky, 1996; Schniederjans, 1993; Schonberger, 1984; Wantuck, 1989.). If additional basic knowledge is thought to be necessary, readers are also encouraged to find one or more of the "B" coded concept books referenced in Appendix I. For those interested in more current literature on introductory JIT concepts and real world JIT applications of principles, reference articles coded "C" in Appendix II.

WHAT ARE "ADVANCED" TOPICS IN JUST-IN-TIME MANAGE-MENT?

Scholars and practitioners are constantly publishing papers on JIT. Some of these papers are theory, some are empirical research, and some are just applications of decades old JIT ideas. While all are a part the total body of knowledge on JIT, the focus of this book is principally on what has been happening with JIT in the more recent years. Specifically, from the mid-1990s to the present, this book examines a collection of JIT research papers focusing on state-of-the-art advances and applications in JIT. What "advanced" topics in JIT management means in this book are the most recent research on JIT methodology, concepts, strategies, and ideas. The research selected to be discussed in this book are from some of the best production, manufacturing, and operations management journals in the world, historically augmented by select older, prior literature in books and articles were appropriate.

This book does not pretend to be comprehensive in presenting JIT advances but instead seeks to cover some of the more emerging topics that scholars and practitioners are now developing. What will be presented in each chapter are an overview of the JIT topic and a discussion of the current research that is related to the topic. While it is not possible to cite all the possible publications that may relate to a particular topic, we have sought to focus on the most recent, significant, and therefore the most "advanced" article publications available in the literature.

WHY LEARN ABOUT ADVANCED TOPICS IN JIT MANAGEMENT?

There are three types of operations in the world: (1) those operations that have not used JIT (at least they think they haven't) and want to, (2) those operations that have been using JIT but want to be brought up to speed on some of the current advances in JIT concepts and methodologies, and (3) the world-class operation's that have been using JIT and want to maintain their leadership by being on the cutting edge of JIT usage. This book can help managers in all three types of operations. The JIT topics selected for this book are presented within the context of the latest literature dealing with that topic. Where possible recent applications of the JIT topic will be described to illustrate how organizations can have success in using JIT.

For the experienced operations manager, it goes without saying that JIT management works. In countless reviews of literature (see Golhar and Stamm, 1991; Sohal, Keller, and Fouad, 1989; Waters-Fuller, 1995; Wilson, 1998) JIT management is consistently a winning philosophy for running operations. But for those who need the scientific justification, this conclusion on JIT has been empirically shown true by academic researchers repeatedly (see McLachlin, 1998; Sakakibara, Flynn and Schroeder, 1993; Sakakibara, Flynn, Schroeder, and Morris, 1997).

THE ORGANIZATION OF THIS BOOK

This book is organized into three parts. In Part I, Chapter 1, "Introduction to Just-In-Time Management," a brief listing of JIT basic principles are presented. This listing is designed to provide a self-check for readers to both realize the breadth of application of JIT principles and possibly find yet unused principles in their operations.

Chapters 2, 3, and 4 represent Part II, "Just-In-Time Quantitative Tools," and provide a discussion on JIT-related quantitative methods. This part of the book is focused on issues currently important in the use of quantitative method- ologies that are appearing in the literature. In Chapter 2, "Economic Justifica- tion and Thresholds for Just-In-Time Implementation," several formulas are presented that can be used to define inventory purchase ordering conditions where JIT can best be applied and where other, more classic methods might be applicable.   Chapter 3, "Topics in JIT Simulation," describes various uses of simulation methods in JIT. In Chapter 4, "Topics in JIT Kanban Analysis," a series of models and conceptual issues on how kanbans are being used in JIT systems are presented.

Part III, "Selected Just-In-Time Topics," consists of four chapters. Chapters 5 and 6 cover a number of different supply chain topics that illustrate the current roles and uses of JIT to maximize distribution and operating production systems. Chapter 7, "Topics in JIT Scheduling," a discussion various JIT scheduling methods and practices are presented. Finally, Chapter 8, "Topics in JIT Human Resource Management," ends the book with a review of human resource man- agement issues and their importance to help make JIT operations successful.

REFERENCES

Adair-Heeley, C. *The Human Side of Just-In-Time: How to Make the Tech- niques Really Work*. New York: American Management Association, 1991.

Ansari, A., and Modarress, B. *Just-In-Time Purchasing*. New York: The Free Press, 1990.

Bare, L. *The Self-instructional Route to Statistical Process Control and Just-In- Time Manufacturing*. Milwaukee, WI: ASQC Quality Press, 1991.

Black, J. T. *The Design of the Factory with a Future*. New York: McGraw-Hill, 1991.

Cheng, T. C. E. and Podolsky, S. *Just-In-Time Manufacturing: An Introduction*. 2nd ed. London: Chapman & Hall, 1996.

Golhar, D. Y., and Stamm, C. L. "The Just-In-Time Philosophy: A Literature Review," *International Journal of Production Research*, 29 (1991), pp. 657-676.

Grieco, P. L. *World Class: Measuring Its Achievement*. New York: P. T. Publi- cations, Inc., 1990.

Hall, R. W. *Implementation of Zero Inventory: Just in Time*. Milwaukee, WI: American Production & Inventory Control Society, Inc., 1986.

Harding, M. *Profitable Purchasing: An Implementation Handbook for Just-in-Time.* New York: Industrial Press, Inc., 1990.

Hirano, H. *JIT Factory Revolution: A Pictorial Guide to Factory Design of the Future.* Cambridge, MA: Productivity Press, 1988.

Hobbs, O. K., "Managing JIT Toward Maturity," *Production and Operations Management Journal*, 38 (1997), pp. 47-50.

Keane, P. T., and King, J. T. *Failing in the Factory: A Shop Floor Perspective on Correcting America's Misunderstanding and Misuse of Just-in-Time.* New York: Brown House Communications, 1991.

Keller, A. Z., and Kazazi, A. "Just-In-Time Manufacturing Systems: A Literature Review," *Industrial Management and Data Systems*, 93 (1993), pp. 3-32.

McLachlin, R. "Management Initiatives and Just-In-Time Manufacturing," *Journal of Operations Management*, 15 (1997), pp. 271-292.

Merli, G. *Total Manufacturing Management: Production Organization for the 1990s.* Cambridge MA: Productivity Press, 1990.

Rutherford, R. D. *Just In Time: Immediate Help for the Time-pressured.* New York: John Wiley & Sons, Inc., 1981.

Sakakibara, S., Flynn, B. B., and Schroeder, R. G. "A Framework and Measurement Instrument for Just-In-Time Manufacturing," *Production and Operations Management*, 2 (1993), pp. 177-194.

Sakakibara, S., Flynn, B. B., Schroeder, R. G., and Morris, W. T. "The Impact of Just-In-Time Manufacturing and Its Infrastructure on Manufacturing Performance," *Management Science*, 43 (1997), pp. 1246-1257.

Schniederjans, M. J. *Topics in Just-In-Time Management.* Boston: Allyn and Bacon, 1993.

Schonberger, R. J. *World Class Manufacturing: The Lessons of Simplicity Applied.* New York: The Free Press. 1986.

Schonberger, R. J. *World Class Manufacturing: The Next Decade: Building Power, Strength, and Value*, New York: The Free Press, 1996.

Schorr, J. E. *Purchasing in the 21ˢᵗ Century: A Guide to State of Art Techniques and Strategies*, New York: John Wiley & Sons, Inc., 1998.

Sohal, A. S., Keller, A. Z., and Fouad, R. H. "A Review of Literature Relating to JIT," *International Journal of Operations and Production Management*, (1989), pp. 15-25.

Vokurka, R. J., and Davis, R. A. "Just-In-Time: The Evolution of a Philosophy," *Production and Operations Management Journal*, 38 (1996), pp. 47-50.

Wantuck, K. A., *Just In Time for America*, Milwaukee, WI: The Forum, Ltd., 1989.

Waters-Fuller, N. "Just-In-Time Purchasing and Supply: A Review of the Literature," *International Journal of Operations and Production Management*, 15 (1995), pp. 220-237.

Wilson, J. M. "A Comparison of the "American System of Manufacturing Circa 1850 with Just in Time Methods," *Journal of Operations Management*, 16 (1998), pp. 77-90.

PART II

JUST-IN-TIME QUANTITATIVE TOOLS

2

ECONOMIC JUSTIFICATION AND THRESHOLDS FOR JUST-IN-TIME IMPLEMENTATION

The purposes of this chapter are to discuss current research on economic methodologies used to justify JIT operations, and demonstrate how these methodologies can be used to determine quantifiable thresholds where the application of JIT maybe questionable. An extension of the economic models to show the superiority of JIT will also be presented.

INTRODUCTION

Why bother to economically justify JIT when virtually all organizations that have applied JIT principles know they work? Because JIT is not always successful and the economic theories can sometimes help identify economic conditions or thresholds under which JIT will and will not work. They can also reveal opportunities that have not yet been fully utilized or considered in any one particular application.

Ocana and Zemel (1996) presented a series of mathematical proofs showing that while JIT can be an optimal policy of inventory management, certain conditions, such as learning from mistakes, must be present. Without those conditions, JIT was shown not to be an optimal policy. Their theoretical work demonstrated that JIT could have limitations and that these limitations are important for JIT to be successfully implemented. This and other very recent literature has shown that JIT has been around long enough for theorists to start re-examining JIT to determine why it doesn't always work.

This chapter presents methodologies in current literature that can be used to define the boundaries or thresholds under which JIT can best be applied. These methodologies can help inventory managers know when and where a JIT approach to inventory is applicable. While it should be mentioned that JIT has a very human side to its successful application (which is discussed in Chapter 8), the focus in this chapter will be on mathematical models that can be easily used to identify critical implementation thresholds.

JUST-IN-TIME AND BASIC ECONOMIC ORDER QUANTITY MODELS

In this section a series of mathematical models are presented that compare a JIT ordering system with an economic order quantity (EOQ) system. These models can be used to compare the cost advantages of either system under differing demand and cost situations. Most important, these models can be used to clearly identify where JIT has cost advantages over an EOQ system, and under what circumstances the EOQ system provides a cost advantage over JIT.

In an article by Fazel (1997), a JIT ordering system is compared with the basic EOQ model originally proposed by Harris (1915). This same basic model is found in virtually every inventory management textbook and in virtually every inventory management software system. The basic EOQ model is an inventory management system that determines the optimal order quantity based on minimizing annual ordering costs (e.g., costs to place the order, ordering technology, forms, etc.), annual holding costs (e.g., costs of material handlers, material handling equipment, taxes, insurance, etc.), and annual purchasing costs (i.e., cost of materials purchased). The total annual cost function for the basic EOQ model used by Fazel (1997) is presented in Table 2.1. By setting this equ-

TABLE 2.1. Total Annual Cost Function for EOQ Model

TC_E = Annual ordering cost + Annual holding cost + Annual purchasing cost, or

$$TC_E = \frac{KD}{Q} + \frac{QH}{2} + P_E D$$

where:
TC_E is the total annual cost using an EOQ order quantity system,
K is the cost of placing an order,
D is annual demand,
Q is order quantity,
H is the annual inventory carrying cost per unit, and
P_E is the purchase cost per unit under an EOQ order quantity system.

---

ation equal to zero and solving for the first derivative with respect to order quantity (i.e., Q), the formula for the optimal order quantity (i.e., $Q*$), is determined as presented in Table 2.2. Incorporating the equation in Table 2.2 into the cost function in Table 2.1, the total annual cost function for an EOQ inventory ordering system is presented in Table 2.3.

To model the JIT ordering costs function, Fazel (1997) simply assumed that cost per unit of ordered inventory under a JIT system (i.e., P_J) would cost more than the cost per unit under the EOQ system (i.e., P_E), because a JIT system re-

TABLE 2.2. Optimal EOQ Model

$$Q* = \sqrt{\frac{2KD}{H}}$$

where:
$Q*$ is the optimal order quantity,
K is the cost of placing an order,
D is annual demand, and
H is the annual inventory carrying cost per unit.

TABLE 2.3. Total Annual Cost Function with Optimal EOQ Model

$$TC_E = \sqrt{2KDH} + P_E D$$

where:
TC_E is the total annual cost at an optimal EOQ order quantity,
K is the cost of placing an order,
D is annual demand,
Q is order quantity,
H is the annual inventory carrying cost per unit, and
P_E is the purchase cost per unit under an EOQ order quantity system.

quires suppliers to perform presetup tasks (i.e., repackaging for unit usage) and store the inventory that would otherwise end up on the floor in the JIT operation. As a result, Fazel's (1997) JIT total annual cost function is simply the product of cost per unit and annual demand, as presented in Table 2.4.

Fazel (1997) then combined the two cost functions in Tables 2.3 and 2.4 to determine the cost difference between the two types of ordering systems. In Table 2.5, the cost difference function (i.e., Z) can be used to easily see when a JIT ordering system is preferable to an EOQ ordering system. By plugging in the necessary demand, price, and cost information, this cost difference function will result in a positive value (meaning that a JIT ordering system is less costly), a zero value (meaning that costs between the two systems are equal), or a negative value (meaning that an EOQ ordering system is less costly).

Fazel (1997) went on to derive several other formulas. The annual demand indifference function presented in Table 2.6 permits the exact demand value or threshold where lesser levels of demand necessitate using a JIT system and greater levels of demand necessitates using an EOQ system. Of particular importance is the maximum JIT purchase price function presented in Table 2.7. This function defines the maximum price that a company should pay per unit of inventory and still use a JIT ordering system.

A Sample Problem for the Fazel Model

To illustrate the informational value of the Fazel models above, let's create a typical sample problem. Suppose we have a single product whose K cost of placing an order is $150, whose D annual demand is variable over a range from 10,000 units to 100,000 units, whose H annual inventory carrying cost per unit is $50, whose P_J product unit price under a JIT system is $24.60, and whose P_E purchase cost per unit under an EOQ order quantity system is $24. This sample problem's data are presented in Table 2.8 for interval values of annual demand of 10,000 units.

TABLE 2.4. Total Annual Cost Function for JIT Model

TC_J = Annual purchasing cost, or

$$TC_J = P_J D$$

where:
TC_J is total annual purchasing cost under a JIT ordering system,
P_J is the product unit price under a JIT system, and
D is annual demand.

TABLE 2.5. Cost Difference Function

Z = (Total Annual Costs Under EOQ) – (Total Annual Costs Under JIT), or

$Z = TC_E - TC_J$, or

$$Z = \sqrt{2KDH} - (P_J - P_E) D$$

where:
Z is the total cost difference between the EOQ and JIT systems,
TC_E is the total annual cost at an optimal EOQ order quantity,
TC_J is total annual purchasing cost under a JIT ordering system,
K is the cost of placing an order,
D is annual demand,
H is the annual inventory carrying cost per unit,
P_J is the product unit price under a JIT system, and
P_E is the purchase cost per unit under an EOQ order quantity system.

Interpretation: If $Z_{qd} > 0$, select JIT system; if $Z_{qd} = 0$, select either system; and if $Z_{qd} < 0$, select EOQ system.

TABLE 2.6. Annual Demand Indifference Function

$$D_I = \frac{2KH}{(P_J - P_E)^2}$$

where:
D_I is the annual demand level where the costs for both systems are equal,
K is the cost of placing an order,
H is the annual inventory carrying cost per unit,
P_J is the product unit price under a JIT system, and
P_E is the purchase cost per unit under an EOQ order quantity system.

TABLE 2.7. Maximum JIT Purchase Price Function

$$P_{JMax} = \sqrt{\frac{2KH}{D}} + P_E$$

where:

P_{JMax} is maximum price per unit that can be paid to keep the JIT system costs under those of the EOQ system,

K is the cost of placing an order,

D is annual demand,

H is the annual inventory carrying cost per unit, and

P_E is the purchase cost per unit under an EOQ order quantity system.

TABLE 2.8. Data for Sample Problem Using Fazel Model

Annual Demand (D)	Holding Cost (H)	Ordering Cost (K)	JIT Price (P_J)	EOQ Price (P_E)
10,000	50	150	24.6	24
20,000	50	150	24.6	24
30,000	50	150	24.6	24
40,000	50	150	24.6	24
50,000	50	150	24.6	24
60,000	50	150	24.6	24
70,000	50	150	24.6	24
80,000	50	150	24.6	24
90,000	50	150	24.6	24
100,000	50	150	24.6	24

Using the formulas, the computed values for the $Q*$ optimal order quantity, the P_{JMax} maximum price per unit that can be paid to keep the JIT system costs under those of the EOQ system, the TC_E is the total annual cost at an optimal EOQ order quantity, the TC_J total annual purchasing cost under a JIT ordering system, and the Z total cost difference between the EOQ and JIT systems are presented in Table 2.9. Note that the cost difference (i.e., Z) becomes negative

TABLE 2.9. Quantity, Cost, and Price Results for Sample Fazel Model Problem

Annual Demand (D)	Order Quantity (Q*)	Max. JIT Price (P_{Jmax})	Total EOQ Cost (TC_E)	Total JIT Cost (TC_J)	Cost Difference (Z)
10,000	244.9490	25.22474	252,247.4	246,000	6,247.449
20,000	346.4102	24.86603	497,320.5	492,000	5,320.508
30,000	424.2641	24.70711	741,213.2	738,000	3,213.203
40,000	489.8979	24.61237	984,494.9	984,000	494.897
50,000	547.7226	24.54772	1,227,389.0	1,230,000	-2,613.870
60,000	600.0000	24.50000	1,470,000.0	1,476,000	-6,000.000
70,000	648.0741	24.46291	1,712,404.0	1,722,000	-9,596.300
80,000	692.8203	24.43301	1,954,641.0	1,968,000	-13,359.000
90,000	734.8469	24.40825	2,196,742.0	2,214,000	-17,257.700
100,000	774.5967	24.38730	2,438,730.0	2,460,000	-21,270.200

at the D annual demand level of 50,000 units on Table 2.2. The actual annual demand level where the two systems have equal cost values can be computed using the formula in Table 2.6. This value turns out to be 41,666.667 units and its computation is presented in Table 2.10. The interpretation of this threshold is that for annual demand levels under 41,666.667 units, a JIT ordering system is less costly. For annual demand levels above 41,666.667 units, an EOQ ordering system is less costly. Also, the P_{Jmax} values in Table 2.9, precisely define the maximum per unit price that the organization can pay for the inventory item and still operate cost effectively under a JIT ordering system. If the per unit prices for the given D annual demand levels fall above the P_{Jmax} values in Table 2.9, then the organization should revert to an EOQ system to reduce total annual inventory costs.

JUST-IN-TIME AND QUANTITY DISCOUNT EOQ MODEL

In a logical extension of the Fazel's (1997) basic EOQ model, Fazel, Fischer, and Gilbert (1998) proposed the same JIT comparative approach, but this time incorporating quantity discounts. In this model, the authors recognize the possibility of availing themselves of quantity discounts on inventory items. The basic premise of this model is the same as the earlier study, but the total annual cost function for the EOQ component is changed to include the possibility of a quantity discount.

As presented in Figure 2.1, there are several unit price levels that have to be considered in a quantity discount model. The same P_J JIT price as previously used is greater than the non-quantity discount price of P^0_E under an EOQ system. There is also an absolute minimum price level (i.e., P^{min}_E) where the full

TABLE 2.10. Sample Problem Demand Indifference Level

$$D_I = \frac{2KH}{(P_J - P_E)^2} = \frac{2(150)(50)}{(24.6 - 24)^2} = 41,666.667 \text{ units}$$

where:
D_I is the annual demand level where the costs for both systems are equal,
K is the cost of placing an order,
H is the annual inventory carrying cost per unit,
P_J is the product unit price under a JIT system, and
P_E is the purchase cost per unit under an EOQ order quantity system.

FIGURE 2.1. Quantity Discount EOQ Variable Cost Function

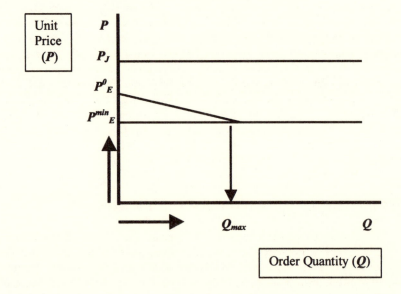

quantity discount is applied. Between prices of $P^0{}_E$ and $P^{min}{}_E$, the discount is variable at a rate of π_E (i.e., a percentage that is used to decrease the initial $P^0{}_E$ price) for each additional unit purchased up to the maximum quantity of Q_{max}. After the quantity of Q_{max} is reached, the discounted price is fixed at $P^{min}{}_E$. Incorporating these components for the range of order quantity purchases from Q

$\leq Q_{max}$, results in the total annual cost function with variable discounts presented in Table 2.11. Fazel, Fischer, and Gilbert (1998) also provided the optimal order quantity for this cost function, as presented in Table 2.12. Note, there is a necessary mathematical condition where $(H - 2\pi_E D) > 0$. This is simply a requirement that prevents the values under the square root sign from becoming negative. It is also a condition that must be met in order for this model to be realistically applied.

Once the quantity of Q_{max} is reached, the discounted price is fixed at $P^{min}{}_E$. This changes the total annual cost function. Incorporating these components for the range of order quantity purchases from $Q > Q_{max}$, results in the total annual cost function with a fixed discounted price as presented in Table 2.13. As noted by Fazel, Fischer, and Gilbert (1998) the optimal order quantity for this cost function is the same as in Table 2.2. This function is further discussed in the next section.

Fazel, Fischer, and Gilbert (1998) developed a number of related models to the cost function in Table 2.11 with the discounted price range occurring in $Q \leq Q_{max}$. These models include the Z_{qd} total cost function for the quantity discount EOQ model with variable discount rate (see Table 2.14). This formula can be used to determine the exact costs between the two systems. Again, it is based on

TABLE 2.11. Total Cost Function for Quantity Discount EOQ Model Where the Order Quantity Discount Is Variable

---

TC_{Eqd} = Annual ordering cost + Annual holding cost + Annual purchasing cost, or

$$TC_{Eqd} = \frac{KD}{Q} + \frac{QH}{2} + (P^0{}_E - \pi_E Q)\, D \quad \text{for } Q \leq Q_{max},$$

where:
TC_{Eqd} is the total annual cost using a quantity discount EOQ order quantity system,
K is the cost of placing an order,
D is annual demand,
Q is order quantity,
H is the annual inventory carrying cost per unit,
$P^0{}_E$ is the initial purchase cost per unit with no quantity discount,
π_E is a constant representing rate at which the price of the item decreases with increases in order quantities, and
Q_{max} is the maximum quantity that can be purchased and still receive a quantity discount whose rate is π_E.

TABLE 2.12. Optimal EOQ Model for the Total Cost Function Where the Discount Is Variable

$$Q^{**} = \sqrt{\frac{2KD}{H - 2\pi_E D}}$$

where:
Q^{**} is the optimal order quantity for a variable discount EOQ ordering system,
K is the cost of placing an order,
D is annual demand,
H is the annual inventory carrying cost per unit, and
π_E is a constant representing rate at which the price of the item decreases with increases in order quantities.

Necessary condition: $(H - 2\pi_E D) > 0$.

TABLE 2.13. Total Cost Function for Quantity Discount EOQ Model Where the Order Quantity Discount Is Fixed

TC_{Eqd} = Annual ordering cost + Annual holding cost
             + Annual purchasing cost, or

$$TC_{Eqd} = \frac{KD}{Q} + \frac{QH}{2} + (P^{min}{}_E) D \quad \text{for } Q > Q_{max}$$

where:
TC_{Eqd} is the total annual cost using a quantity discount EOQ order quantity system,
K is the cost of placing an order,
D is annual demand,
Q is order quantity,
H is the annual inventory carrying cost per unit,
$P^{min}{}_E$ is fixed purchase cost per unit when an order quantity exceeds a certain maximum level of Q_{max}, and
Q_{max} is the maximum quantity that can be purchased and still receive a quantity discount whose rate is π_E.

TABLE 2.14. Cost Difference Function for the Fazel, Fischer, and Gilbert Quantity Discount EOQ Model Ordering Systems with Variable Discounts

Z_{qd} = (Total Annual Costs Under Quantity Discount EOQ)
 − (Total Annual Costs Under JIT), or

$Z_{qd} = TC_E - TC_J$, or

$$Z_{qd} = kD \sqrt{\frac{H - 2\pi_E D}{2KD}} + \frac{H}{2} \sqrt{\frac{2KD}{H - 2\pi_E D}}$$

$$+ \left[(P^0_E - \pi_E) \sqrt{\frac{2KD}{H - 2\pi_E D}} \right] D - P_J D$$

where:
Z_{qd} is the total cost difference between the quantity discount EOQ and JIT systems,
TC_E is the total annual cost at an optimal quantity discount EOQ order quantity,
TC_J is total annual purchasing cost under a JIT ordering system,
K is the cost of placing an order,
D is annual demand,
H is the annual inventory carrying cost per unit,
π_E is a constant representing rate at which the price of the item decreases with increases in order quantities,
P^0_E is the initial purchase cost per unit with no quantity discount, and
P_J is the product unit price under a JIT system.

Necessary condition: $(H - 2\pi_E D) > 0$.

Interpretation: If $Z_{qd} > 0$, select JIT system; if $Z_{qd} = 0$, select either system; and if $Z_{qd} < 0$, select EOQ system.

the JIT annual cost function in Tables of 2.4 and 2.5. Fazel, Fischer, and Gilbert (1998) also presented the formula for the D_I annual demand level where the costs for both systems are equal (see Table 2.15). This formula can be used to find the threshold point where JIT system is the most cost effective for ordering inventory.

TABLE 2.15. Annual Demand Indifference Function for Variable Discount Fazel, Fischer and Gilbert Ordering Systems

---

$$D_I = \frac{2KH}{(P_J - P^0_E)^2 + 4K\pi_E}$$

where:
D_I is the annual demand level where the costs for both systems are equal,
K is the cost of placing an order,
H is the annual inventory carrying cost per unit,
P_J is the product unit price under a JIT system,
P^0_E is the purchase cost per unit under an EOQ order quantity system, and
π_E is a constant representing rate at which the price of the item decreases with
    increases in order quantities.

A Sample Problem for the Fazel, Fischer, and Gilbert Models

To illustrate the informational value of the Fazel, Fischer, and Gilbert models above, let's create another typical sample problem. Suppose we have a single product whose K cost of placing an order is $50, whose D annual demand is variable over a range from 10,000 units to 100,000 units, whose H annual inventory carrying cost per unit is $300, whose P_J product unit price under a JIT system is $25.60, whose initial P^0_E purchase cost per unit under a quantity discount EOQ order quantity system is $24, whose final P^{min}_E purchase cost per unit under a quantity discount EOQ order quantity system is $22.40, and whose π_E is a constant 0.0001 representing a rate at which the price of the item decreases with unit purchases up to Q_{max}. Also, in this problem Q_{max} has to be 16,000 units because the difference between the P^0_E price of $24 and the P^{min}_E price of $22.40 allows a range of $1.6 at a rate of 0.0001(i.e., 1.6/0.0001 = 16,000 unit range). This sample problem's data is presented in Table 2.16 for interval values of annual demand of 10,000 units.

The resulting computations using the formulas in Tables 2.11, 2.12, 2.13, and 2.14 for the order quantities and cost information are presented in Table 2.17. It can be seen that the cost difference that favors a JIT system occurs before the D annual demand level of 20,000 units. The exact D_I annual demand level can be determined using the formula in Table 2.15. Plugging the given values of this sample problem into the formula and simplifying, the resulting threshold value for a D_I annual demand is 11,627.91 units (see Table 2.18 for

TABLE 2.16. Data for Quantity Discount Fazel, Fischer, and Gilbert Model Sample Problem

Annual Demand (D)	Holding Cost (H)	Ordering Cost (K)	JIT Price (P_J)	EOQ Price (P^0_E)	Discount Rate (π_E)
10,000	300	50	25.6	24	0.0001
20,000	300	50	25.6	24	0.0001
30,000	300	50	25.6	24	0.0001
40,000	300	50	25.6	24	0.0001
50,000	300	50	25.6	24	0.0001
60,000	300	50	25.6	24	0.0001
70,000	300	50	25.6	24	0.0001
80,000	300	50	25.6	24	0.0001
90,000	300	50	25.6	24	0.0001
100,000	300	50	25.6	24	0.0001

TABLE 2.17. Quantity and Cost Results for Quantity Discount Fazel, Fischer, and Gilbert Sample Problem

Annual Demand (D)	Order Quantity (Q^{**})	Total EOQ Cost (TC_E)	Total JIT Cost (TC_J)	Cost Difference (Z_{qd})
10,000	57.9284	257,262.7	256,000	1,262.6
20,000	82.1994	504,331.1	512,000	-7,668.9
30,000	101.0153	749,698.5	768,000	-1,8301.5
40,000	117.0411	994,176.0	1,024,000	-29,824.0
50,000	131.3064	1,238,079.0	1,280,000	-41,921.1
60,000	144.3376	1,481,569.0	1,536,000	-54,430.8
70,000	156.4466	1,724,744.0	1,792,000	-67,256.3
80,000	167.8363	1,967,666.0	2,048,000	-80,334.5
90,000	178.6474	2,210,379.0	2,304,000	-93,621.4
100,000	188.9822	2,452,915.0	2,560,000	-107,085.0

TABLE 2.18. Fazel, Fischer, and Gilbert Sample Problem Demand Indifference Level

---

$$D_I = \frac{2KH}{(P_J - P^0_E)^2 + 4K\pi_E} = \frac{2(50)(300)}{(25.6 - 24)^2 + 4(50)(0.0001)} = 11{,}627.91 \text{ units}$$

where:
D_I is the annual demand level where the costs for both systems are equal,
K is the cost of placing an order,
H is the annual inventory carrying cost per unit,
P_J is the product unit price under a JIT system, and
P^0_E is the purchase cost per unit under an EOQ order quantity system.

---

computations). This value falls within the 16,000 unit limitation and does not violate the necessary condition where $(H - 2\pi_E D) > 0$ (i.e., $300 - 2(0.0001)(11{,}628) > 0$). The interpretation of the D_I threshold is that for values of D below D_I a JIT ordering system should be used to minimize costs, and for annual demand values above D_I the quantity discount EOQ ordering system should be used.

A LOGICAL EXTENTION OF THE COMPARATIVE EOQ AND JIT MODELS

It would appear from the Fazel (1997) and Fazel, Fischer and Gilbert (1998) models that JIT ordering systems are fairly limited to low annual demand in inventory items (see Z and Z_{qd} columns in Tables 2.9 and 2.17, respectively). If, we accept the assumptions under which Fazel (1997) and Fazel, Fischer and Gilbert (1998) models operate, then from an economic basis, JIT ordering systems are limited. A closer examination by others (Schniederjans and Qing, 1999) has revealed a logical extension of the previous models that point to a very different conclusion.

In Schniederjans and Qing (1999) a revision of the Fazel, Fischer and Gilbert (1998) model is proposed. This revision seeks to revise the JIT total annual cost component to recognize the realistic advantage of physical plant space reductions caused by a reduction in incoming, work-in-process, and finished goods inventory. Research on JIT systems have consistently documented the inevitable reduction in facility square feet when adopting a JIT inventory ordering system. JIT experts, such as Schonberger (1982, pp. 121-122) and Wantuck (1989, p. 16), and countless JIT implementations (see Chase, Aquilano and Jacobs, 1998a; 1998b; Hay 1988, pp. 22-23; Jones, 1991; Stasey and McNair,

1990, Chapter 13; and Voss 1990, p. 330). These research studies suggest JIT was responsible for reductions ranging from 30 percent of a plant up to 80 percent of the total space of an existing plant.

Schniederjans and Qing (1999) included the square foot reduction that is attributable to JIT by adjusting the TC_J component to be reduced by the cost reduction created when reducing the plant size. The annual facility cost reduction is found by multiplying the cost per square foot of facility space times the number of square feet saved using a JIT ordering system. This cost component is presented in the revised total annual cost function under a JIT ordering system as presented in Table 2.19. The value of C is computed from overhead costs (i.e., material handlers saved, facility insurance, facility taxes, etc.). The value of N is best determined when a change from an EOQ system is made to a JIT system. Based on the prior research on JIT implementations, the estimation of N is not difficult. Including this revised component into the cost difference function Z_{qd} in Table 2.14, produces a newly revised cost difference function, Z_{qdR}, as presented in Table 2.20.

A Sample Problem for the Schniederjans and Qing Model

To illustrate the impact of adding the annual facility cost reduction as suggested by Schniederjans and Qing (1999), let's again create a typical sample problem. Suppose we have a single product whose K cost of placing an order is $50, whose D annual demand is variable over a range from 10,000 units to 100,000 units, whose H annual inventory carrying cost per unit is $300, whose P_J product unit price under a JIT system is $26, whose initial P^0_E purchase cost per unit under an EOQ order quantity system is $24, whose π_E is a constant 0.0001 representing a rate at which the price of the item decreases with unit purchases up to Q_{max}. In addition, the annual cost to own and maintain a square foot of facility C is $100, and N number of square feet saved by initially adopting a JIT system is 125,000. This 125,000 figure assumes an average plant of 500,000 square feet and only a 25% reduction due to JIT implementation. This sample problem's data is presented in Tables 2.21 and 2.22 for interval values of annual demand of 10,000 units.

TABLE 2.19. Revised JIT Model Component

TC_{JR} = Annual purchasing cost – Annual facility cost reduction, or

$$TC_{JR} = P_J D - CN$$

where:
TC_{JR} is the revised total annual purchasing cost under a JIT ordering system,
P_J is the product unit price under a JIT system,
D is annual demand,
C is the annual cost to own and maintain a square foot of facility, and
N is the number of square feet saved by initially adopting a JIT system.

TABLE 2.20.   Cost Difference Function for Revised Quantity Discount Schniederjans and Qing Ordering Systems

Z_{qdR} = (Total Annual Costs Under Quantity Discount EOQ)
       − (Total Annual Costs Under JIT), or

$Z_{qdR} = TC_E - TC_{JR}$, or

$$Z_{qdR} = kD \sqrt{\frac{H - 2\pi_E D}{2KD}} + \frac{H}{2}\sqrt{\frac{2KD}{H - 2\pi_E D}}$$

$$+ \left[(P^0_E - \pi_E)\sqrt{\frac{2KD}{H - 2\pi_E D}} \right] D - (P_J D - CN)$$

where:
Z_{qdR} is the revised total cost difference between the EOQ and JIT systems,
TC_E is the total annual cost at an optimal quantity discount EOQ order quantity,
TC_J is total annual purchasing cost under a JIT ordering system,
K is the cost of placing an order,
D is annual demand,
H is the annual inventory carrying cost per unit,
π_E is a constant representing rate at which the price of the item decreases with
      increases in order quantities,
P^0_E is the initial purchase cost per unit with no quantity discount,
P_J is the product unit price under a JIT system
C is the annual cost to own and maintain a square foot of facility, and
N is the number of square feet saved by initially adopting a JIT system.

Necessary condition: $(H - 2\pi_E D) > 0$.

Interpretation: If $Z_{qdR} > 0$, select JIT system; if $Z_{qdR} = 0$, select either system; and if $Z_{qdR} < 0$, select EOQ system.

The resulting computations using the formulas in Tables 2.13 and 2.20 for the order quantities and cost information are presented in Table 2.22. It can be seen that at every level of D annual demand the cost difference favors a JIT system. The Schniederjans and Qing (1999) model suggest that when cost factors, like JIT reducing plant physical facility costs, are included in the total ann-

TABLE 2.21. Data for Revised Quantity Discount Schniederjans and Qing Model Sample Problem

Annual Demand (D)	Holding Cost (H)	Ordering Cost (K)	JIT Price (P_J)	EOQ Price (P^0_E)	Discount Rate (π_E)
10,000	300	50	26	24	0.0001
11,000	300	50	26	24	0.0001
12,000	300	50	26	24	0.0001
13,000	300	50	26	24	0.0001
14,000	300	50	26	24	0.0001
15,000	300	50	26	24	0.0001
16,000	300	50	26	24	0.0001
17,000	300	50	26	24	0.0001
18,000	300	50	26	24	0.0001
19,000	300	50	26	24	0.0001
20,000	300	50	26	24	0.0001

TABLE 2.22. Additional Data, Quantity, and Cost Results for Revised Quantity Discount Schniederjans and Qing Model Sample Problem

Annual Demand (D)	Cost Per Sq. Foot (C)	JIT Reduced Sq. Footage (N)	Order Quantity (Q^{**})	Cost Difference (Z_{qdR})
10,000	100	125,000	57.93	12,501,262.68
11,000	100	125,000	60.78	12,500,499.17
12,000	100	125,000	63.50	12,499,697.62
13,000	100	125,000	66.12	12,498,862.65
14,000	100	125,000	68.63	12,497,998.04
15,000	100	125,000	71.07	12,497,106.87
16,000	100	125,000	73.42	12,496,191.74
17,000	100	125,000	75.71	12,495,254.84
18,000	100	125,000	77.93	12,494,298.05
19,000	100	125,000	80.09	12,493,322.98
20,000	100	125,000	82.20	12,492,331.05

ual cost functions, JIT ordering systems are clearly the least costly system to use. Indeed, the magnitudes of the Z_{qdR} cost differences are so great, that computing the cost indifference point becomes a redundancy. Moreover, the extremely large values of D where the quantity discount EOQ model might start showing a cost preference, will in virtually every case, violate the necessary condition (i.e., $H - 2\pi_E D > 0$), which is required for its use.

Does the results of the Schniederjans and Qing (1999) model suggest that JIT should always be used and quantity discount EOQ models should never be considered in inventory planning? No each model has its own potential area of application. As Gupta and Kini (1995) carefully have pointed out, JIT and EOQ quantity discounts can be successfully taken advantage of to reduce total inventory costs. In situations where plants do not experience or can not take advantage of their square footage reductions under a JIT system, the Fazel (1997) and Fazel, Fischer, and Gilbert (1998) models may provide very useful thresholds defining JIT system feasibility. On the other hand, Fazel, Fischer and Gilbert (1998) and Gupta and Kini (1995) have both pointed out that the theoretical nature of economic costing models do not consider many of the other advantages that a JIT system can offer users. Such advantages as flexibility, quality, process improvement and a host of other advantages are commonplace in JIT systems and should be included in economic modeling of any JIT system.

SUMMARY

This chapter presented a very recently developed series of EOQ and JIT models useful in finding thresholds under which each model was most cost effective. The basic EOQ and quantity discount EOQ models were both compared with JIT ordering systems. The models were shown to be useful in identifying thresholds were JIT ordering systems were more cost effective than EOQ systems and where EOQ ordering systems were less costly than JIT systems. In addition, this chapter discussed a logical extension of the prior research by considering additional cost information on facility space reduction usually observed in JIT implementations. By including the reduced square footage cost advantage to the JIT side of the comparison model, a JIT system was found to be preferable to an EOQ system, for any annual demand level.

This chapter's content is designed to help inventory managers understand the advantages, as well as the disadvantages of using a JIT inventory ordering system. Representing recently advanced methodology, the models in this chapter will continue to undergo development in the literature as academic and professionals add to their JIT knowledge. One obvious limitation in the models presented in this chapter was their deterministic nature. That is, the parameters in the models where required to be fixed values. As every practicing manager knows, everything in a JIT environment is dynamic and always changing. To help deal with the uncertainties and variability's in a JIT environment, simulation methods can be used to model JIT systems. This subject is covered in Chapter 3.

REFERENCES

Chase, R. B., Aquilano, N. J., and Jacobs, F. R. *JIT at Federal Signal*, Vol. 4, Irwin Operations Management Video Series. Boston: Irwin/McGraw-Hill, (1998a).

Chase, R. B., Aquilano, N. J., and Jacobs, F. R. *Tristate Converting to JIT: Parts 1& 2*, Vol. 5, Irwin Operations Management Video Series. Boston: Irwin/McGraw-Hill, (1998b).

Fazel, F. "A Comparative Analysis of Inventory Costs of JIT and EOQ," *International Journal of Physical Distribution and Logistics*. 27, (1997), pp. 496-505.

Fazel, F., Fischer, K. P., and Gilbert, E. W. "JIT Purchasing vs. EOQ with a Price Discount: An Analytical Comparison of Inventory Costs," *International Journal of Production Economics,* 54 (1998), pp. 101-109.

Gupta, O., and Kini, R. B. "Is Price-Quantity Discount Dead in a Just-In-Time Environment?," *International Journal of Operations and Production Management*, 15 (1995), pp. 261-270.

Harris, F. W. *Operations and Cost-Factory Management Series*. Chicago: A. W. Shaw Co., 1915, chapter 4.

Hay, E. J. *The JIT Breakthrough: Implementing the New Manufacturing Basics*. New York: John Wiley & Sons, 1988.

Jones, D. J. "JIT and the EOQ Model: Odd Couples No More," *Management Accounting*, 72 (1991), pp. 54-57.

Ocana, C., and Zemel, E. "Learning From Mistakes: A Note on Just-In-Time Systems," *Operations Research*, 44 (1996), pp. 206-214.

Schniederjans, M. J., and Qing, C. "A Note on JIT Purchasing vs. EOQ With a Price Discount: An Analytical Comparison of Inventory Costs," Midwest Decision Sciences Institute Meeting, Springfield, IL, (1999). (Soon to be published in the *International Journal of Production Economics*.)

Schonberger, R. J. *Japanese Manufacturing Techniques*. New York: The Free Press, 1982.

Stasey, R., and C. J. McNair, C. J. *Crossroads: A JIT Success Story*. Homewood, IL: Dow Jones-Irwin, 1990.

Voss, C. A. *International Trends in Manufacturing Technology: JIT Manufacture*. London: Springer-Verlag, 1990.

Wantuck, K. A. *Just In Time for America*. Milwaukee, WI: The Forum, 1989.

3

TOPICS IN JUST-IN-TIME SIMULATION

This chapter provides an overview of simulation methodology and explain how is it used in JIT systems. A brief tutorial on simulation is provided, as well as a review of the use of simulation in JIT, and a state-of-the-art assessment of current simulation topics in JIT.

WHAT IS SIMULATION?

For years people have used simulation to approximate the operation of systems. Simulations have been used to train people for flight, train employees for complex operations, and approximate the operation of a process or system. In general, the term *simulation* refers to the imitation of a real world system over time (Banks, Carson, and Nelson, 1996).

Simulation has been used to analyze a variety of systems. For the evaluation of the performance of manufacturing systems, simulation has become the dominant methodology. Special modeling tools and simulation software have been developed just for the application to various manufacturing systems. Other production and operating systems that have been modeled include transportation systems, computer/communications systems, military systems, agricultural systems, hospitals, and virtually any other business-related process. Crucial to the discussion of this chapter, simulation can be used to measure JIT performance and operational procedures.

WHY USE SIMULATION?

One of the primary advantages of simulation is that it can give an accurate approximation of a real world system. Analytical modeling techniques such as linear programming are able to generate an optimal solution to a problem, but have difficulty solving problems which are probabilistic in nature. Also, when analytical models have a large number analytic functions (and some very complex stochastic functions) built into a model they become extremely difficult to solve. The construction of simulation models allows users to analyze these complex systems. Once the simulation model is developed and validated, a model can be used to answer a wide variety of "what if" questions about the real world system (Banks, Carson, and Nelson, 1996). Perhaps one of the most important reasons why simulation is used is that it permits changes to an operating system without risk to actual system (i.e., the changes can be simulated in order to predict their impact on system performance).

Simulation is one of the most frequently used system analysis methods. There are a number of reasons for using simulation. Simulation can be used to analyze models of any level of complexity. The sophistication of the model is limited only by the ability of the modeler to represent the detailed workings of the system, and the computer's capacity to run the simulation program. The modelers primary interest might be to use the model in order find the design that maximizes one or more performance measures or simple study the behavior of the system.

Often the modeling effort in itself is useful. The process of model development requires the system to be studied and understood. The study frequently uncovers problems that were unknown or not understood previously. Relative to other methodologies, simulation, can carry more credibility with decision makers. Animated simulations include a visual representation of the model, and can be used to demonstrate that the model actually approximates system performance. In addition, the visual representation is often an important point that can sell an idea to management or other key decision makers (Rohrer, 1996). Finally, the same set of simulation methods can be used to analyze a stochastic system, regardless of structure or complexity.

SIMULATION SOFTWARE

Simulation applications are generally accomplished with the use of specially developed simulation software. There are several alternatives for building simulation models. They can be classified into three major categories: general programming languages, general simulation languages, and special purpose simulation tools. Any problem can be simulated by any of these tools. Given these alternatives, how should a user select a tool? To answer this question, consider the advantages and disadvantages of each class of simulation languages.

1. *Generalized programming languages.* Examples of these are *C, Basic* and *Fortran.* The primary advantages of these general simulation languages are that they provide

maximum flexibility and control over output format. This flexibility allows the user to analyze virtually any system. The primary disadvantage is that the user must code and develop the entire model.

2. *General simulation languages.* Examples of these include *SLAM, SIMAN*, and *GPSS/H/PC*. The advantages of general simulation languages over general computer languages are that model development is easier, model development and refinement are faster, and models are easier to understand. The primary disadvantage is that the model developer loses some flexibility over generalized programming languages. In addition, the modeler must learn a new programming language to use the tool.

3. *Special purpose simulation languages.* Examples of these include *PROMODEL, WITNESS* and *ARENA*. The advantages of special purpose simulation tools over general purpose languages are they are easier to learn, generally very little programming knowledge is needed, and model output are very easy to understand. In addition, the graphical component in these programs gives another presentation medium. The primary disadvantages are the models are relatively inflexible and no real knowledge of simulation is needed to operate the system. This causes potential problems because the modeler does not have to understand the system to build the simulation model. In these instances certain key operational variables will often be left out of the model giving decision makers poor information on which to base their decisions.

There is an ever growing number of simulation systems available for both the PC computer platform and mainframe computer platform. For an extensive list of simulation software and the program characteristics, see Swain (1995).

STEPS TO A SIMULATION STUDY

There are many different ways in which to conduct a good simulation study. Figure 3.1 shows the steps that compose a typical simulation study and their relationships. Similar figures and discussion can be found in other sources (see Banks, Carson, and Nelson, 1996; Law and Kelton, 1991). Some studies may not necessarily contain all of the steps in the order stated below and still other studies may have additional steps, which are not depicted in the diagram. The steps outlined here are a suggested guideline for conducting a simulation study.

1. *Formulate the problem and plan the study.* Every study should begin with a clear statement of the problem and plans for how the study is to be conducted, and the criterion of how to evaluate the alternative system designs.

2. *Model conceptualization.* This is also known as the "art" of modeling. Each problem has a unique set of criterion, which are important to the solution of the problem. It is the modeler's job to determine which of those characteristics that are important to the development of the best model. The best method is to start with a simple model and build toward a complex system. The model can be developed either graphically or in logic format to define the components, descriptive variables, and interactions (logic) that constitute the system.

3. *Collect the data.* Data collection is highly integrated with model conceptualization and will determine the data collected. It is important to remember at this stage that data collection takes a large portion of the total time to conduct the simulation study.

4. *Construct simulation model.* A decision must be made a to use a general programming language, general simulation language, or special purpose language to model the situation. Once the appropriate tool is chosen the user must write and code the simulation model.

5. *Verification.* Involves debugging the model and establishing that the computer program is working as intended. A number of techniques exist to help in this stage of the process including using animation in verificaton and using an interactive tracer and debugger.

6. *Validation.* Establish that a desired accuracy or correspondence exists between the simulation model and the real system. There are many techniques for achieving this, but a common approach for establishing the validity/credibility of the model is for the analyst to perform a structured walk through of the conceptual model to an audience of key people. This helps to ensure that the model's assumptions are correct, complete, and consistent.

7. *Experimental design.* Design an experiment that will yield the desired information and determine how each of the test runs specified in the experimental design are to be executed. In order for the simulation to be statistically precise and free of bias the modeler should make sure run times are sufficiently long and account for the startup conditions of the system.

8. *Make production runs.* Perform production runs as specified in step 7.

9. *Analyze output data.* Draw inferences from the data generated by the simulation. The output data from the production runs are used to construct numeric estimates of the desired measures of performance.

10. *Document and implement results.* Simulation models are often used for more than one application, good documentation is very important and should include an assumptions document, documentation of the program, and reporting and summarized results of the study.

Simulation provides an additional tool in which to base decisions. The ability of simulation models to approximate real systems is what makes it useful in JIT research.

JIT AND SIMULATION

As mentioned previously simulation can be used to analyze many different types of systems. Simulation is primarily used in manufacturing research to explain characteristics of operating systems, which may be too difficult to model analytically. Common manufacturing elements analyzed using simulation models include scheduling systems, inventory control policies, warehousing control features, production planning, control elements, and logistic systems. Common to all of these systems is the stochastic nature of the system. The stochastic element in system is what makes them appropriate to use simulation.

Some of the stochastic system topics have been researched in a JIT environment. Most of the research has focused on the differences between "push" and "pull" manufacturing environments. Previous areas of JIT simulation research have focused on the following areas: kanbans, implementation in manufacturing, cellular manufacturing, and purchasing. Only the most recent previous simulation research conducted in JIT is discussed. For an in-depth discuss-

FIGURE 3.1. Steps in a Simulation Study

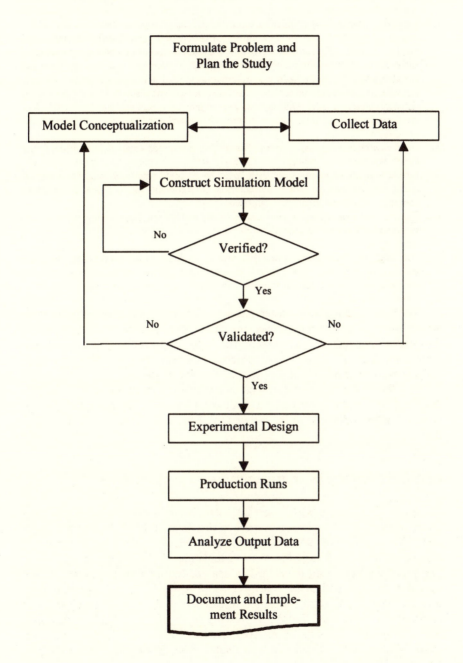

ion of early simulation studies in JIT see, Golhar and Stamm (1991), Keller and Kazazi (1993), and Sohal, Keller, and Kazazi (1988).

1. *Kanbans*. The bulk of simulation studies in JIT have been conducted in kanban research. In most JIT systems, a majority of production control and material movement is controlled by kanbans. Previous simulation studies of kanban systems typically consider the environment where demand is uncertain, processing times are variable and machine breakdowns are random (Golhar and Stamm, 1991). The primary focus was to determine the optimal number of kanbans to reduce the amount of inventory in the system. Dispatching and assignment rules have also been a focus of previous kanban research. For a continued discussion of kanban related research, see Chapter 4.
2. *Implementation in manufacturing*. Simulation on implementation in manufacturing issues has been researched since the early 1980s. Primarily they have been concerned with the differences between "push" and "pull" type manufacturing environments. Considerations for manufacturing environment, random demand, random stoppages and breakdowns have all been researched.
3. *Cellular manufacturing*. The objective of the research in this area has been the development of U-shaped assembly lines that have increased efficiency and performance.
4. *Purchasing*. The previous simulation studies in this area have examined the feasibility of distribution centers in JIT environments. Other simulation studies have compared current JIT purchasing practices to EOQ type environments.

The result of simulation research has enabled many organizations to achieve higher rates of production, lower inventory levels, increase manufacturing throughput, and reduce cycle time. The individual results of the simulation models are generally specific to a particular situation, however, they give some insights as to the general problems that exist in JIT production.

The focus of much of the simulation continues to be in the analysis of production planning and control in stochastic environments. The remaining portion of this chapter is devoted to the discussion of the current topics in JIT simulation.

CURRENT TOPICS IN SIMULATION AND JUST-IN-TIME

The focus of simulation is no longer just manufacturing systems but extends to a much wider range of topics. Advances in the capabilities of the newer simulation software programs have helped to allow users to model a wider variety of systems without much experience with simulation. Indeed, specialized simulation programs have been developed to analyze transportation systems, service operation systems, and health care systems. The advent of the various simulation packages has allowed simulation researchers to analyze a wider variety of performance measures. Table 3.1 presents some of the current performance measures used in JIT simulation research (see Ardalan, 1997; Kern and Wei, 1996; Savasar, 1996; Takahashi, and Izumi, 1997; Takahashi, Nakamura, and Ohasmia, 1996). While this list is certainly not comprehensive, it does outline the wide range of subjects that can be analyzed using simulation. In addition, it

TABLE 3.1. Performance Measures Used in JIT Research

- Average finished goods inventory
- Total units of sales lost
- Schedule instability
- Throughput rate
- Station utilization
- Total WIP levels
- Average time to fill a customer order
- Total inventory
- Average customer waiting time
- Number of full containers in the system
- Average cycle time
- Minimum number of kanbans required
- Machine downtime
- Batch size
- Distribution time
- Total supply-chain stock
- Customer service rates

helps to point out the ability of simulation to give measurements of stochastic variables.

In the next four sections of this chapter some of the most current areas of JIT simulation research are highlighted. Areas included in the discussion are production planning and control, sequencing/scheduling rules, kanbans, and JIT logistics/supply-chain design. (The focus of the kanban and sequencing/scheduling material will be on the nature of what was simulated and not the specifics of the models. These topics are covered in greater detail in Chapters 4 and 7, respectively.)

SIMULATION AND PRODUCTION PLANNING AND CONTROL

While new simulation software has been developed to simulate a wide variety of topics, production planning and control continues to receive great attention in simulation literature. Various elements of *production planning* and *control* (PP&C) have been simulated in the past few years. Current analysis has given insights into maintenance policy selection, performance measures, alternative system configurations, part routings, and many other PP&C elements.

Maintenance Policies

One of the critical elements in a smooth running JIT production system is the reliability and dependability of production equipment. One of the most effective ways to keep production equipment operating efficiently is to develop a *preventive maintenance* (PM) program. PM allows an organization to routinely check their production equipment for worn out parts and service the existing parts to keep the machinery running smoothly. While the adoption of PM seems to be intuitively obvious, the actual consequences incurred by a production system have not previously been examined. Savasar (1997b) presented a simulation model that analyzed the effects of different maintenance policies on performance measures of JIT production systems under different operating conditions. The simulation model developed was for a line consisting of five stations in series and operated according to demand from the last station. It showed that the introduction of preventive maintenance to the production and assembly machines increases line performance under all conditions. The simulation model is specific to this particular example, but the research results indicated that PM programs can require a fair amount of resources decrease downtime to the system effectively.

Production Strategies

One of the biggest advantages to using simulation, as a means of systems analysis, is that it allows the user to test new strategies for systems. Within the context of PP&C the types of programs tested have included production batch size, kanban strategies, unit load size, and many other production variables.

Several case studies have shown how specific production strategies can be tested using simulation. For example, Savasar (1997a) built a simulation model that tested various production strategies on an electronic assembly line. The model scheduled weekly demands for production control boards from its final board inventory. In this system, demand triggers production at the first station by signaling for replenishment of a quantity equal to the demand. The assembly operations themselves were performed using a push system of production control. This model illustrated a hybrid push/pull production strategy (i.e., demand is triggered or pulled and the assembly operations are scheduled via a push strategy). The objective of the simulation analysis was to determine the minimum number of batches in the system to meet a certain percentage of demand on time. The specific results are not important as they are case specific, but they do indicate that both pull and push strategies can operate in the same production environment. Each element is better suited for a particular function in the production system.

In another case study, Estrada, Villalobos, and Roderick (1997) investigated some of the theoretical and practical issues for the introduction of JIT techniques into an automotive wire-harness assembly line. Issues explored include the number of production kanbans and the unit load size to be used for every pair of subassembly stations in a production line. Four performance criteria were used to evaluate the strategies: throughput, work-in-progress inventory, lead time, and subassembly equipment utilization.   The performance measures used in this study are specific to the nature of this problem. The simulation tool allowed the modeler to measure the efficiency and effectiveness of system performance.

Software Development

General programming languages give the experienced programmers the ability to develop applications, which can analyze complex problems. The development of these specific programs has most frequently occurred in the area of production planning and control. For example, Wang and Xu (1997), examined the performance of the hybrid push-pull production control strategy. They developed a strategy simulation software for flow shop manufacturing systems. It used a structured model to describe manufacturing and assembling processes of any production line. By using the model, the material flow controlled by different strategies can be simulated by different descriptive equations. This particular application allows a user to analyze a wide variety of production strategies for JIT flow shop environments.

SIMULATION AND KANBAN

Since kanbans represent the control of material movement in production systems this continues to be a likely candidate for simulation. Most production environments are highly stochastic and more often than not demand is uncertain. Simulation provides the perfect tool to analyze these production environments. Since topical area of kanbans is covered in greater detail in Chapter 4, this section highlights the nature of the problems simulated and provide illustrations of those environments. In this section we will discuss how simulation is used to model kanban production environments, and used with other quantitative methods (i.e., genetic and evolutionary algorithms, and metamodeling) to deal with kanban behavior.

Production Environments

The purpose for simulation in kanban research is to help determine kanban policies in various production settings. Many analytical models have been developed to handle the deterministic one and two card kanban systems. The use of simulation helps develop kanban policies for the more difficult multiple product, multiple production line settings. Baykoc and Erol (1998) provided a typical JIT production environment simulation situation. Typical of the JIT production environment, a collection of work stations and production areas (i.e., assembly area) are interconnected as presented in Figure 3.2. The objective for the simulator is to model each of these components and the stochastic properties. The particular manufacturing environment in Figure 3.2 can be characterized by having multiproducts and multistages. Using this type of framework, Atwater and Chakrovarty (1996) compared the output performance of pull production systems with different interstation storage capacities under various levels of total

FIGURE 3.2. Example of a Typical Simulated JIT Manufacturing Environment

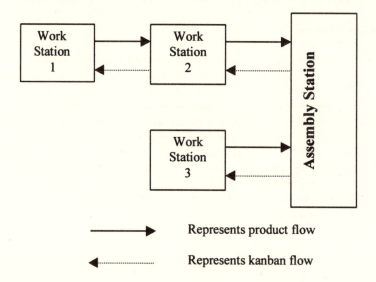

Represents product flow

Represents kanban flow

inventory in the system, station production variations, and station downtime. They found that pull systems with larger interstation storage capacities yield significantly higher output at all levels of inventory in the system, regardless of station variations and station downtimes. They also showed that lines using a strict kanban-based pull system can increase output while decreasing WIP inventory simply by allowing the inventory to flow freely between stations.

Genetic and Evolutionary Algorithms

One of the more recent techniques developed to generate near optimal solutions to complex problems are the *evolutionary search* (ES) heuristic and *genetic algorithms* (GA). These algorithms take complex problem sets and search for near optimal solutions. They are computer-based generators that create and examine thousands of potential solutions to a problem.   However, the initial solution may only be a local optimum (i.e., not the global optimal solution). The algorithms, having generated this near optimal point, continue too ever so slightly manipulate the problem set (i.e., genetically evolve) to recalculate the solution to the problem.   This process of evolution is repeated an indefinite amount of times until a near and often optimal solution is created. The use of GA and ES in combination with simulation has provided a very powerful tool to determine the optimal number of kanbans in various production environments.

Several studies describe the use of evolution strategies integrated with a simulation model, which includes a stochastic processes of a manufacturing system, to solve the kanban sizing problem. The ES heuristic determines the minimum number of kanbans and corresponding production trigger values required to meet demand. The newly created search heuristics are based on both the ES heuristic and the classic kanban sizing equation used by Toyota (Boyden, Hall, and Usher 1996; Hall, Bowden, and Usher, 1996). The combination of using simulation with other tools enables users to build models which are capable at generating multiple solutions simultaneously.

Metamodeling

Metamodeling is a simulation technique aimed at developing a simple mathematical model to explain a more complex simulation model (see Banks, Carson, and Nelson 1996). The modeler generally will begin by simulating a large experimental design in which the dependent variable is explained by a large number of independent variables. In most cases the functional relationship between the independent and dependent variable is not known. The analyst using this approach chooses an appropriate method for estimating the unknown parameters in the relationship. Regression analysis is one of the most common methods to explain the relationships in metamodeling.

Aytug, Dogan, and Bezmez (1996) describes a method to determine the number of kanbans in a pull production system by using simulation metamodeling. The method is demonstrated on a two-card kanban controlled manufacturing system. Through metamodelling, a relationship between the number of kanbans and the average time to fill a customer order is determined. This relationship is later used in a model to determine the number of kanbans while minimizing cost. The nice feature of metamodels is that the relationship generated is in functional format allowing the analyst to quickly estimate the impact of changing one or more of the independent variables.

SIMULATION AND SEQUENCING/SCHEDULING RULES

Another area, that has received a lot of attention in recent years has been sequencing/scheduling rules for machine operations. The dynamic nature of sequencing rules in JIT makes it a perfect candidate for simulation. The majority of the sequencing environments are stochastic in nature and makes the use of analytical models very difficult. One of the areas, which have received a great deal of attention, is sequencing rules for mixed model assembly lines.

Organizations using assembly lines are required to turn out a higher variety of products, higher quality products, and all with shorter lead times (see Sumichrast and Clayton, 1996). The evolution from Ford's single-product, no-variety production line is the *mixed-model assembly line*. Complicating the problem is that in JIT environments the assembly line is paced because of the production due date constraints. In general the sequencing methods in this environment seek to account for the variation in production and make production level or uniform. The prominent mixed-model assembly line goals for JIT systems have been broadly classified into two categories: the subassembly usage smoothing goal and the product load-smoothing goal (see Monden, 1993). The first is to ensure a constant rate of usage into the line and the second to reduce the occurrences of line stoppages. Aigbedo and Monden (1996), through a very detailed set of simulation experiments, demonstrated using the subassembly usage goal in two level systems provides a better sequence that can be developed over other heuristics.

New findings suggest that sequencing methods that have been developed to achieve a constant use of components do not actually do so very well in a paced mixed-model assembly environment. While some of the common sequencing rules like goal chasing, time spread, and batch do not ensure a constant rate of production, simulation studies have found them to be good heuristics because they promote efficient assembly (Sumichrast and Clayton 1996). (These and other scheduling/sequence-related research are discussed in Chapter 7.)

SIMULATION AND LOGISTICS/SUPPLY-CHAIN DESIGN

The focus of most of prior simulation analysis in JIT has been toward the operation and control of the production system. Recently the boundaries of application are expanding. Topics such as systems design, logistics analysis, and even integrated supply channel analysis have been simulated.

System-Wide Analysis

In JIT we often describe the overall goal to permit smooth operation of the system is to eliminate waste from the system. Waste can take many different forms such as excess inventory, rework, scrap, physical waste, and idle operation time. Lost in many of the microlevel analyses presented earlier are the ramifications of making those changes on the rest of the system. Often positive changes in one aspect of the organization can lead to negative changes in a different aspect of the company. Simulation allows the modeler to take into account the majority of those corporate wide aspects. Smet and Gelders (1997) described how the effect of the elimination of waste on performance measures such as lead time and overtime costs can be estimated using logistical simulation. The study uses simulation to conduct a sensitivity analysis which evaluates the effects of different types of waste, such as machine breakdowns, inadequate buffer size and production time variance on operational performance. In general, the aggressive removal of waste from a system leads toward a more efficient system in terms of reduced cycle times, increased throughput and lower operating costs.

Supply-Chain Design and Analysis

As competition grows ever more fierce, organizations are relying more and more on the capabilities of the supply-chain rather than the capabilities of an individual firm. Although each supply-chain organization may regard their role as satisfying the demand of its immediate customer, the more holistic view suggests that each is just a part of a wider supply-chain system whose role is to satisfy end customer demand (Berry and Naim, 1996).

Models are now being built that describe the dynamic implications of various supply-chain redesign strategies adopted by large organizations (Berry and Naim, 1996; Lee, Padmanabahn, and Whang, 1997). The models reveal a number of features common to all supply-chains:

1. Demand in the marketplace become delayed and amplified moving upstream in a supply-chain.
2. Supply-chain designs tend to magnify the effects of changing market demand.
3. Supply-chain designs can introduce random variations due to information lags often mistaken as seasonal or periodic variations.
4. Attempts to reduce poor supply-chain behavior can magnify the problem. This often results because of the lack of information shared in the system.
5. Poor forecasting and other demand assessment techniques help to magnify the effect.

These phenomena are often referred to as the "bullwhip effect" (see Chapter 5). The focus of the simulation model in case is to gain insights as to the appropriate supply-chain design so as to minimize the effects of the bullwhip effect. The simulation studies concluded that the following elements helped to reduce the effects of these phenomena: the introduction of the JIT philosophy in manufacturing plants, the development of global materials planning systems which attain visibility of total supply-chain stock, a strategic supplier sourcing policy, and the bypassing of the distribution network so as to directly interface with the customer (Berry and Naim, 1996; Lee, Padmanabahn, and Whang, 1997).

SUMMARY

The methodological process of simulation analysis was briefly described in this chapter. After a discussion of recent research showing how simulation analysis is used in topical areas of JIT management, including production planning and control, kanbans, sequencing/scheduling rules, and logistics and supply-chain design is provided.

This chapter has tried to make the point that simulation can provide a powerful tool to analyze systems of any level of complexity. As described in this chapter, there are many types of simulation software programs available to perform these structured analyses. Analysts can use very specialized programs to evaluate specific problems or they can develop their own application using general programming languages to analyze any system.

The use of simulation in JIT research has allowed modelers to solve complex problems. Some of these simulation studies have examined the impact of JIT and its usage of kanbans within the organization. The next chapter explore kanban analysis.

REFERENCES

Aigbedo, H., and Monden, Y. "A Simulation Analysis for Two-Level Sequence-Scheduling for Just-In-Time (JIT) Mixed-Model Assembly Lines," *International Journal of Production Research*, 34 (1996), pp. 3107-3124.

Ardalan, A. "Analysis of Local Decision Rules in a Dual-Kanban Flow Shop," *Decision Sciences*, 28 (1997), pp. 195-211.

Atwater, J. B., and Chakravorty, S. S. "The Impact of Restricting the Flow of Inventory in Serial Production Systems," *International Journal of Production Research*, 34 (1996), pp. 2657-2669.

Aytug, H., Dogan, C. A., and Bezmez, G. "Determining the Number of Kanbans: A Simulation Metamodeling Approach," *Simulation*, 67 (1996), pp. 23-32.

Banks, J., Carson, J. S., and Nelson, B. L. *Discrete Event Simulation, Second Edition.* Upper Saddle River, NJ: Prentice-Hall, 1996.

Baykoc, O. F., and Erol, S. "Simulation Modeling and Analysis of a JIT Production System," *International Journal of Production Economics*, 55 (1998), pp. 203-212.

Berry, D., and Naim, M. M. "Quantifying the Relative Improvements of Redesign Strategies in a PC Supply-chain," *International Journal of Production Economics*, 46-47 (1996), pp. 181-196.

Boyden, R. O., Hall, J. D., and Usher, J. M. "Integration of Evolutionary Programming and Simulation to Optimize a Pull Production System," *Computers & Industrial Engineering*, 31 (1996), pp. 217-220.

Estrada, F., Villalobos, J. R., and Roderick, L. "Evaluation of Just-In-Time Alternatives in the Electric Wire-Harness Industry," *International Journal of Production Research*, 35 (1997), pp. 1993-2008.

Golhar, D. Y., and Stamm, C. L. "The Just-In-Time Philosophy: A Literature Review," *International Journal of Production Research*, 29 (1991), pp. 657-676.

Hall, J. D., Bowden, R. O., and Usher, J. M. "Using Evolution Strategies and Simulation to Optimize a Pull Production System," *Journal of Materials Processing Technology*, 61 (1996), pp. 47-52.

Hum, S. H., and Lee, C. K. "JIT Scheduling Rules: A Simulation Evaluation," *Omega, International Journal of Management Science*, 26 (1998), pp. 381-395.

Kellaer, A. Z. and Kazazi, A. "Just-In-Time Manufacturing: A Literature Review," *Industrial Management and Data Systems*, 93 (1993), pp. 2-32.

Kern, G. M. and Wei, J.C. "Master Production Rescheduling Policy in Capacity-Constrained Just-In-Time Make-To-Stock Environments," *Decision Sciences*, 27 (1996), pp. 365-387.

Law, A. M.,and Kelton, D. W. *Simulation Modeling and Analysis,* 2nd ed. New York: McGraw-Hill, 1991.

Lee, H. L., Padmanabhan, V., and Whang, S. "The Bullwhip Effect in Supply-chains," *Sloan Management Review*, 86 (1997), pp. 93-102.

Monden, Y. *Toyota Production System: An Integrated Approach to Production Management*. Atlanta, GA: Industrial Engineering and Management Press, 1993.

Rohrer, M. "Visualization and its Importance in Manufacturing Simulation," *Industrial Management*, May/June (1996), pp. 15-18.

Savasar, M. "Effect of Kanban Withdrawal Policies and Other Factors on the Performance of JIT Systems-A Simulation Study," *International Journal of Production Research*, 34 (1996), pp. 2879-2899.

Savasar, M. "Simulation Analysis of a Push/Pull System for an Electronic Assembly Line," *International Journal of Production Economics*, 51 (1997a), pp. 205-214.

Savasar, M. "Simulation Analysis of Maintenance Policies in Just-In-Time Production Systems," *International Journal of Production Research*, 17 (1997b), pp. 256-266.

Smet, R., and Gelders, L. "Evaluation of the Role of Waste in a Truck Manufacturing Using Simulation," *Journal of Intelligent Manufacturing*, 8 (1997), pp. 449-458.

Sumichrast, R. T., and Clayton, E. R. "Evaluating Sequences for Paced, Mixed-Model Assembly Lines with JIT Component Fabrication," *International Journal of Production Research*, 34 (1996), pp. 3125-3143.

Sohal, A. S., Keller, A. Z., and Found, R. H. "A Review of Literature Relating to JIT," *International Journal of Operations and Production Management*, 9 (1989), pp. 15-25.

Swain, J. J. "Simulation Survey: Tools For Process Understanding and Improvement," *OR/MS Today* (August 1995), pp. 64-79.

Takahashi, K., Nakamura, N., and Izumi, M. "Concurrent Ordering in JIT Production Systems," *International Journal of Operations and Production*, 17 (1997), pp. 267-290.

Takahashi, K., Nakamura, N., and Ohashi, K. "Order Release in JIT Production Systems: A Simulation Study," *Simulation*, 66 (1996), pp. 75-87.

Wang, D., and Xu, C. G. "Hybrid Push/Pull Production Control Strategy Simulation and its Applications," *Production Planning and Control*, 8 (1997), pp. 142-151.

4

Topics in Kanban Analysis

The intent of this chapter is to explain the use of kanbans in JIT production. After a brief introduction to kanbans, a discussion on current issues in kanban control and design is presented.

WHAT IS KANBAN?

In general manufacturing systems can be classified into "push" or "pull" systems. The difference between the two types of systems are that pull systems initiate production in response to current demand and push systems initiate production in response to expected or forecasted demand (see Karmarker, 1989). *Kanbans* and their use in JIT are examples of pull production systems.

Kanban systems are a means of controlling inventory and production scheduling in JIT environments. The Japanese translation of kanban is card or signpost (Ohno, 1988). The kanban system uses cards or signals to govern the flow of materials though the manufacturing facility. One of the first kanban systems was developed by the Toyota Motor Corporation and was implemented in the 1950s and 1960s (Im, 1989; Monden, 1983). Scheduling in JIT certainly is much more complex than the use of kanban systems. However, this chapter is dedicated to the discussion of kanban systems and development. (For a more in-depth discussion on JIT scheduling refer to Chapter 7.)

To better illustrate the kanban discipline, it is useful to demonstrate the differences between push and pull environments. In general, in a pull system, production is a result from customer orders entering the system. Kanban signals are sent upstream from the final work station to begin production of the next part. As each station completes their work the signal indicates to the upstream station to move parts downstream. The downstream station is pulling parts from the previous work center. In a push system, work releases are triggered by a production schedule. As soon as work is completed at a work center the part is pushed to the next work station. Figure 4.1 graphically illustrates the difference between push and pull manufacturing environments.

FIGURE 4.1. Comparison between Push and Pull Environments

a. Push Manufacturing Environment

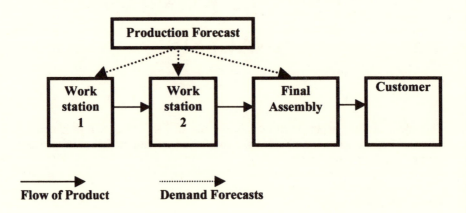

b. Pull Manufacturing Environment

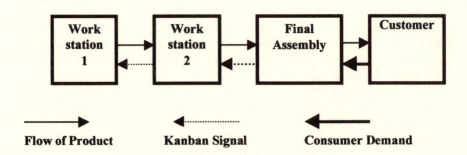

TYPES OF KANBAN CARDS

Kanban cards are used in a JIT environment to control the movement of material, authorization of production, or the ordering of materials from vendors. Each activity requires a different type of kanban card. The three basic forms of kanban cards are the product kanban, vendor kanban, and the material movement (convehance) kanban (Schniederjans, 1993; Singh, 1996).

1. *Product kanban.* The primary function of the product kanban is to signal that production can begin at a preceding station. The product kanban contains information about the product name, part number, stock location, and the materials required for production. Figure 4.2(a) is a graphical illustration of a hypothetical product kanban.
2. *Vendor kanban.* The vendor card is sent to suppliers ordering a specified amount of materials or supplies. The material contained on this card is related to the vendor's product identification, name, amounts, and often times due dates.
3. *Material movement (convehance) kanban.* This kanban is used to authorize the movement of parts and supplies from one work center to another or from raw stock to a workstation. Contained on the convehance kanban are information related to the product name, product identification, lot size and routing process, name, and destination of the subsequent process. Figure 4.2(b) is a graphical illustration of a hypothetical convehance kanban.

While there are the three basic forms of kanbans cards are not the only way to signal production, movement, or ordering of parts (Chase, Aquilano, and Jacobs, 1998, pp. 329-332). Kanbans themselves come in many shapes and sizes. To convey the movement or production of material, many organizations use a *tote* (i.e., a tray or cardboard box used to temporarily hold inventory) or container system. Once a tote is emptied it signals to either begin new production or replenish the tote contents, and return the contents to the work station. This system offers some benefits over the traditional card approach in that it is highly visible and in some cases the materials handler does not have to count material, just simply fill the container. Other kanbans include marking spaces on the floor and alternative vendor cards include using railroad car hoppers and empty semitrailers.

DUAL- AND SINGLE-CARD SYSEMS

There are not many differences between single-card and dual-card kanban systems. *Single-card kanban systems* are generally used to convey the movement of material within a system. Single card systems work very effectively in situations where work stations are spatially very close to each other and there is some excess inventory in the system available for pickup (Schniederjans, 1993). *Dual-card kanban systems* are very similar but they use one or more of the kanban cards listed earlier. The original Toyota kanban system is an example of a dual-card kanban system.

FIGURE 4.2. Examples of Production and Convehance Kanban Cards

a. Production Kanban Card

Product Name Identification # Stock Location	• Number of parts needed • Materials required • Preceding work center • Next work center

b. Convehance Kanban Card

Product Name Identification # Stock Location	• Number of parts needed • Station to deliver the materials

Like any managerial procedure, a kanban system will operate under certain conditions better than under others. There are many elements that allow kanban systems to operate smoothly and efficiently. Some of the general operational rules for kanbans include to never move material without a kanban, only the exact amount of material should be moved, only good parts are moved along in the process, and production should be uniform and constant (see Monden 1983; Schniederjans 1993; Singh 1996).

KANBAN METHODOLOGIES

One of the major issues surrounding the differences in the operating environments is the number of kanbans required at each work station. The original formula used to determine the number of kanbans that are used to support production for the Toyota system can be represented by the basic formula in Table 4.1 (see Wantuck, 1989, p. 276).

TABLE 4.1. Basic Formula for Determining the Number of Kanbans

---

$$n_p = \frac{(d)(t)(1 + e)}{c}$$

where:
n_p is the number of kanbans required for each work station,
d is the average daily demand,
t is the average setup time, and
c is the capacity of the containers in the kanban systems.

This model represents a simple way to determine the number of kanbans needed for production. However, there are many issues that this simple model does not answer. Issues such as the number of kanbans needed in an uncertain environment and how to use kanban in operational environments, which are different than the Toyota production system. The remainder of this chapter is devoted to the discussion of recent advances in kanbans and some of the new techniques surrounding kanban research.

OPTIMIZATION MODELS IN KANBAN

There are many subjects in kanban analysis that can be optimized. Our discussion here will be focused on the optimization of the number of kanbans in the system.

Number of Kanban Cards

One of the big issues for any organization that is considering a switch to a kanban control mechanisms is how many kanban cards to have in the system. Simply because the amount of kanbans in the system will determine the throughput rates and the amount of inventory in the system. If the production environment is stable (i.e., uniform production rates, constant lead times, etc.), the model above works quite efficiently. Other factors to consider in determining the optimal number of kanban cards include operation environment, machine breakdowns, demand uncertainty, and set-up frequency and times (Philipoom, Rees, and Taylor, 1996; Price and Gravel, 1995; Yan, 1995; Ohno, Nakashima, and Kojima, 1995). This section presents two methods for determining optimal number of kanbans with one or more of the variables listed above.

One of critical uncertainty components that impact the number of kanbans required in a system is the issue of variable lead times and uncertain demand. Rees, Philipoom, and Taylor (1997) presented an iterative procedure that can calculate the *probability mass function* (PMF) for the number of kanbans under the conditions of variable lead time and uncertain demand. Singh (1996) presented a logical extension to the model by determining the optimal number of kanbans in this environment based on holding and shortage costs of inventory. In this model two scenarios are presented.

Scenario 1. When the number of kanbans circulating in the system is less than the actual requirements a shortage cost is incurred.

Scenario 2. When the number of kanbans circulating in the system is greater than the actual requirements a holding cost is incurred.

From these two scenarios the total expected cost function is derived (see Table 4.2). Once the total cost equation is derived the optimal number of kanbans in the system can be determined (see Table 4.3).

A Sample Problem for the Singh Model

Suppose we are given the information in Table 4.4 about a production facility. The probability information represents the PMF for number of kanbans based on existing lead times and forecasts. Also, the shortage costs and holding costs per container are $100 and $50, respectively. From this information we can determine the optimal number of kanbans for this system. The solution to the problem is as follows:

1. Using the probabilistic model determine the value for $S/(S+H)$, which for this example is $100/(100 + 50) = 0.6667$.
2. From the PMF determine the value for n that gives $P(n - 1) < 0.6667 < P(n)$. For this particular problem that number is 3.

This model provides the ability to account for several dynamic components to determine the optimal number of kanbans. It accounts for forecasted demand, variable lead times, holding and shortage costs. It does not account for other elements such as machine breakdowns.

Two other concerns for organizations using kanbans are equipment failure and variable demand. The question as to how to set the optimal number of kanbans in the system while maintaining low levels of inventory and operation efficiency is essential for many operations. Yan (1995) presented a model for determining optimal number of kanbans in an environment where demand and machine breakdowns are uncertain. While an optimal number of kanbans can be determined for simple problems under these conditions, the problem becomes too difficult when large complex problems are introduced. To handle problems of this complexity, heuristics are used to give close approximations of an optimal solution.

In the model, Yan (1995) presented an iterative heuristic for approximating the near optimal number of kanbans in a system with one machine and one part type (see Table 4.5). The approach used in the development of this model uses

TABLE 4.2. Total Expected Cost Function for Number of Kanbans

---

a. Scenario 1: Number of kanbans circulating is less than actual requirements:

$$\text{Expected Holding Costs} = H \sum_{y=0}^{n} (n-y)\,p(y)$$

b. Scenario 2: Number of kanbans circulating is greater than actual requirements:

$$\text{Expected Shortage Costs} = S \sum_{y=n+1}^{\infty} (y-n)\,p(y)$$

We can represent total expected cost by:

$$TC(n) = \left[H \sum_{y=0}^{n} (n-y)\,p(y) \right] + \left[S \sum_{y=n+1}^{\infty} (y-n)\,p(y) \right]$$

where:
$TC(n)$ is the total cost of the number of kanbans in a system,
$p(y)$ is the probability mass function for number of kanbans (see Rees,
Philipoom, and Taylor, 1997),
n is the number of kanbans in the system,
y is the number of kanbans needed,
H is the holding cost per container at each work station,
S is the shortage cost per container at each work station, and
∞ is an expression representing infinity.

TABLE 4.3. Optimal Number of Kanbans Based on Probabilistic Costs

---

$$P(n-1) < \frac{S}{S+H} < P(n)$$

where:
$P(n)$ is the cumulative distribution function of n,
H is the holding cost per container at each work station, and
S is the shortage cost per container at each work station.

TABLE 4.4. Kanbans Data for Sample Problem

Probability	Number of Kanbans
0.1	0
0.2	1
0.4	2
0.3	3
0.1	4

TABLE 4.5. Yan Algorithm for Approximating Number of Kanbans with Stochastic Demand and General Machine Breakdowns

Begin

For: $i = 1$ to M
          Set $D = 0$
                    For: $j = 1$ to T_n
                    If $x(j) > 0$;
                    $D = D + 1$
                    $K(i+1) = K(i) - \varepsilon T_n^{-1}(u*D - b*(T_n-D))$
End

where:
M = some predetermined number of trials to simulate,
D = counter,
$x(j)$ = gradient estimate of production function at time j,
$K(i+1)$ = number of kanbans at time $i + 1$,
$K(i)$ = number of kanbans at time i,
T_n = length of time horizon at time n,
ε = a constant,
u = unit inventory costs, and
b = unit backorder costs.

the technique of perturbation analysis. Essentially, *perturbation analysis* enables modelers to estimate how sensitive a performance measure is respect to system parameters. The perturbation analysis calculates the gradient of the performance measures with respect to the system parameters.

In this particular problem the gradient estimate is based on machine availability, production rate, demand rate, and machine capacity level. The solution to this algorithm yields a value for the number of kanbans in a system and minimizes the long-run average cost function subject to the conditions of stochastic demand and machine failures. Simulation results demonstrate indicate the heuristic holds well under most conditions of uncertain demand and general machine breakdowns.

The number of kanban cards in a system is not the only parameter that is of concern when developing a kanban control system. Other system parameters that must be optimized in order to minimize total system and annual costs include number of machines to devote to a particular operation and the kanban container size. Berkley (1996) examined the effect of container size on average inventory and service levels in a two-card kanban system processing multiple parts. The results show that smaller container sizes lead to smaller amounts of average inventory and better service levels in most circumstances. For most of the optimization models the solution of the mathematical model is not practical and certainly could not be implemented in a normal time frame. However the development of these models is what allows users to set the system parameters for simulation models. These simulation models often lead to the development of heuristics as presented in Table 4.5.

ALTERNATIVE FORMS OF KANBAN

As with any production planning and control system, situations exist where use of the original system as it was designed is neither feasible nor appropriate. For most organizations the initial method gets tailored and modified to fit their particular production environment. Monden (1983, p. 64) describes several conditions that make it difficult to use a kanban system. These conditions are: (1) work orders with short production runs, (2) work orders with significant set-up times, (3) systems with significant scrap loss, and (4) stochastic demand patterns. Obviously a lack of smooth and uniform production is nearly impossible under these conditions. The next two section alternative forms of kanban systems are suggested.

Concurrent Ordering System

The intent of the kanban system or any JIT production control method is to produce product only when needed, the goal being to keep WIP levels at a minimum while maintaining acceptable levels of throughput and keeping total annual costs low. Some problems can arise in the use of kanban systems when orders are delayed due to variances in production time. For example, a delay can occur when a kanban is sent to a preceding stage and not enough parts are available to satisfy demand, the order must be delayed. This in turn results in work stations and people going idle for large periods of time.

Takahashi, Nakamura, and Izumi (1997) presented an alternative to the traditional kanban systems to handle the problems associated with order delay. In their model they suggest when incoming demand enters the system the production orders are released to all preceding stages in the production process. In this respect their process is a *concurrent ordering process*. Therefore, in a concurrent ordering process when a final product is delivered the kanban signal releases an order to produce a new product, and all associated subassemblies and parts. This varies significantly from the traditional kanban process, which is a serial ordering process (i.e., kanbans send work orders back to only the immediate stage in the production process).

What are the associated trade-offs related to this new concurrent ordering process? The trade-off is essentially between WIP and cycle time. In the concurrent ordering process, delay time and therefore system cycle time is reduced, but WIP increases. Under the traditional kanban method WIP is reduced, but system cycletime is increased. Under both conditions throughput for the system is the same. For more insights into this phenomenon, we can refer to Little's Law:

$$\text{Average Throughput} = \frac{\text{Average WIP}}{\text{Average System Time}}$$

Under steady-state conditions, the average WIP of a system is equal to the product of average throughput and average system time. In the concurrent ordering process the amount of WIP in the system was sacrificed for reducing system cycle time. Takahashi, Nakamura, and Izumi (1997) concluded that the concurrent ordering systems are more effective than the traditional kanban systems in environments where there is variability in the product lead times and production rates. This is just one alternative to the traditional kanban systems. The CONWIP system presents another variation of the kanban discipline.

CONWIP System

Constant work-in-process (CONWIP) is a variation of traditional kanban systems (Hopp and Spearman, 1996; Spearman, Woodruff, and Hopp, 1990). Instead of sending the kanban cards to the preceding station, CONWIP sends production cards back to the start of the entire production system. In general, production kanbans are used to signal production of a specific part. CONWIP production cards are assigned to a production line but are not a part specific system (see Figure 4.3 for a comparison between kanban and CONWIP). Production cards are assigned to a production line, part numbers are assigned from the production backlog. As work is needed for the first production center, a production card is assigned to the first part in the queue in which raw materials are available.

By using the CONWIP system, organizations literally set a target WIP level for a production line. When work is removed from the final stage in the system, the CONWIP card is sent back to the beginning of the line authorizing the production of a new product, meaning that a CONWIP system will not begin a job unless there is available space in the production sequence. This is significantly

FIGURE 4.3. Comparison between Kanban and CONWIP

a. Kanban system

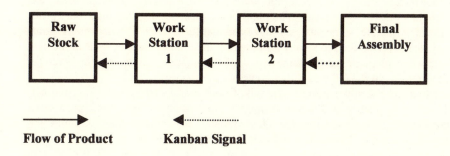

b. CONWIP System

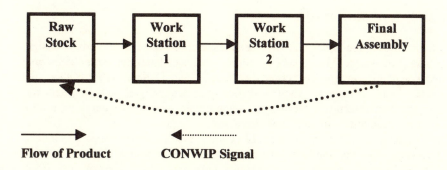

different than the traditional kanban system, which controls the specific sequence of material flow through each work station. CONWIP does not govern the flow of material within the production line, but only the amount of WIP that is allowed in the system. Once the job is released into the system, the jobs are pushed through the production process.

While kanban was specifically designed for a uniform production, level demand environment, CONWIP works in more general production environments. Since CONWIP only controls the amount of WIP in the system, product variation, set-up times, and uncertain demand has minimal impact of the operation of the system.

CONWIP was originally designed for systems in which standard container sizes were used in a single production line on a first come, first served basis. The associated issues related to CONWIP control are very similar to that for kanbans. Critical elements to consider in the design of the CONWIP system include: the number of cards in the system, target WIP and production quotas, and the signal mechanism to be used in the system.

CONWIP has proven to be very effective for some organizations. Roderick, Toland, and Rodriguez (1994) examined the impact of the CONWIP order release strategy versus a pure *materials requirements planning* (MRP) environment at Westinghouse. The plant is a typical small batch electronic products manufacturer. The analysis suggested that the CONWIP order release strategy was a very effective means for reducing the total amount of WIP in the system while maintaining the same levels of throughput. However, for other organizations, CONWIP was shown to be not as effective as the traditional kanban systems (Bonvik, Couch, and Gershwin, 1997; Gstetttner, 1996). Suggesting production managers must consider that every production environment has its own unique characteristics that will largely determine the system to adopt. Instances exist that may require a hybrid of both systems to permit the operation of smooth running production system (Bonvik, Couch, and Gershwin, 1997; Duenyas, 1994; Duenyas and Keblis, 1995). To determine which system is the most effective managers must consider the operational characteristics of the environment.

KANBAN DESIGN ISSUES

The majority of the issues that have been researched regarding the proper design of a kanban system involve simple models, such as single-line, single-item, and single-machine centers (see Frein, DiMascolo, and Dallery, 1995; Keller and Kazzi, 1993). In contrast, the real difficulty in designing a kanban system for the typical manufacturing facility is when there are several product lines, multiple production lines, and uncertain demands.

Important to the design of a kanban control systems is the impact of random processing times and random demand arrivals on the impact of system performance of kanban-controlled environments (Andijani 1997; Baycok and Erol, 1998). To measure the impact of these sources of variation we can measure system performance in terms of output rate, waiting time, and station utilization. Simulation models can be used to examine the impact of these conditions on system performance.

One of the principle methods to develop an effective design for a kanban system is through the use of simulation models. These models allow us to experiment with various system parameters. Andijani (1997) presented a generalized design rule on how to go about constructing simulation models to test kanban system design. The problem in using simulation to test any design is deciding how to go about developing models that will be useful for in comparing various alternatives. For example, consider an environment where X machines are available and Y kanban cards to allocate. To effectively test this situation all the feasible solutions we would have to use a full factorial design. That is, the total number of simulations needed to test the entire spectrum can be calculated by $(Y - 1 / X - 1)$. For relatively small problems we can effectively test every combination to see which of the alternatives is the best. For large systems the amount of alternatives to test can be very large. Andijani (1997) developed a heuristic to use to determine which simulation alternatives to test that will maximize system time and yet minimize system time (see Table 4.6). This heuristic gives an organization a method to effectively develop a kanban system in a realistic time frame.

More important for purposes of this chapter is the impact of various system conditions on system performance and the ramifications they have on kanban design. Some general design issues can be observed with respect to the impact of variable processing times and variable demand rates (Baycok and Erol, 1998). The system performance measures of interest are: total throughput, waiting time and station utilization.

1. *Total throughput.* In general, an increase in the number of kanbans will increase total throughput. However, there are threshold levels when an increase in total number of kanbans has no effect on throughput and only saturates the system with inventory. On the other hand, an increase in system *coefficient of variation* (CV) (i.e., mean/standard deviation) will reduce total throughput. In general, the system is much more sensitive to changes in CV as opposed to changes in the number of kanbans.
2. *Waiting time.* For this performance measure, the system is much more sensitive to the changes in number of kanbans. Changes in CV have very little impact on waiting time when compared to number of kanbans.
3. *Station Utilization.* Simply the utilization increases as the amount of kanbans increase and the CV decreases.

TABLE 4.6. General Design Rule Steps for Kanban Allocation

Consider the general situation in a serial production line where:

Y kanban cards are to be distributed over X machine centers

1. Allocate a kanban at each work station and $(Y - X + 1)$ kanbans at the last station.
2. Record as one of the alternatives to test.
3. Move one kanban from the last work station to the previous work station.
4. Repeat Steps 2 and 3 until there is one kanban left at the last work station. At this point we will have one kanban at each work station.
5. Since downstream work stations account for less average system time, begin moving and distributing one kanban from a downstream work station towards an upstream work station.
6. After each move, record as one of the alternatives to test.
7. Stop if a uniform allocation in the interior work stations has been achieved.
8. Simulate and use results in designing the appropriate system.

For the optimal design of any JIT kanban system depends greatly on the organization removing excess waste and variation from the system. One of the methods in which organizations have used to improve the control of waste and general planning is through the use of modern information technology tools.

INFORMATION TECHNOLOGY AND KANBAN

The majority of U.S. manufactures use computer support and other information technologies to develop and control production schedules. The complexity of many production environments requires the use of computer integrated technology to track and record the use of materials and inventory. Kanban systems, on the other hand, were developed as a noncomputerized scheduling control mechanisms. The literal translation of kanban revolves around visual and not computerized technology.

Recent inventions in computerized technology have made the use of computer support and kanban systems very effective (Childe, 1997). User-friendly software and on-screen technology have permitted many organizations to develop their kanban systems online. The proper use of *information technology* in kanban system development provides some advantages to the traditional kanban control mechanism.

1.   It provides users with up to date information about job status. This is useful in the coordination of the flow of materials between work centers.
2.   It provides users up-to-date information on the on-hand inventory. This helps operators decide which job is to go next in the production sequence.
3.   It accurately records the receipt and release of materials in the production system.
4.   Kanban amounts are easily adjusted and updated.
5.   Organizations do not have to worry about replacing lost kanban cards.
6.   Works easily with barcode systems.
7.   Works effectively in systems that are spread out spatially. The electronic transmission of the kanban places no limit on the size of the production system.

Many organizations must use electronic kanbans to permit smooth operation of the system. For example, organizations whose primary product is related to information processing makes the use of electronic kanban essential. Credit card processing and payroll organizations are just a few who use electronic kanbans. Prior to the actual submission of the data, these organizations perform many activities that allow for the rapid processing of materials once the monthly data is submitted. Once the preprocessing activities are performed the activity center sends an electronic kanban to the organization to have them submit the appropriate data.

Electronic Data Interchange (EDI) and the Internet have provided organizations with a mechanism for transmitting production and inventory data between organizations. They have provided a logical medium for the use of vendor authorization kanban cards (Srinivasan, Kekre, and Mukhopadhyay, 1994). For a more complete description of EDI and other interfirm data transmission methods refer to Chapter 5.

As more firms begin to rely on the supplier and contractor capabilities to effectively compete in the marketplace, electronic data transmission will become more widespread. The control of inventory of systems that are spread out over many organizations will become more difficult and the use of electronic kanbans will become more prevalent.

SUMMARY

Variation can exist in demand, product variety, processing times and in many other facets of the organization. The initial development of a kanban system at the Toyota production facility dealt with a rigid environment where production was fairly constant and the product variety was low. This chapter presented models that help deal with the design and control of kanban systems in environments where there can exists a great deal of variation.

This chapter initially began with a model that discusses the optimal number of kanbans in an environment where the lead times and demand are stochastic and inventory shortage and holding costs occur. This chapter also described CONWIP, a system designed to help develop kanban systems where product variety and processing time are variable. Finally, a discussion was presented regarding the design issues a manager must consider in developing a kanban system and what new technologies that exist to help in kanban control.

The purpose of this chapter was to help production control managers understand the complexities surrounding JIT kanban systems. But before WIP can move within an organization using a kanban system, it must be delivered to the organization. The next chapter provides an introduction to the logistic and distribution systems responsible for delivery to and from JIT operations.

REFERENCES

Andijani, A. "Trade-off between Maximizing Throughput and Minimizing System Time in Kanban Systems," *International Journal of Operations & Production Management*, 17 (1997), pp. 429-445.

Baykoc, O. F., and Erol, S. "Simulation Modeling and Analysis of a JIT Production System," *International Journal of Production Economics*, 55 (1998), pp. 203-212.

Berkley, B. J. "A Simulation Study of Container Size in Two-Card Kanban Systems," *International Journal of Production Research*, 34 (1996), pp. 3417-3445.

Bonvik, A. M., Couch, C. E., and Gershwin, S. B. "A Comparison of Production-line Control Mechanisms," *International Journal of Production Research*, 35 (1997), pp. 789-804.

Chase, R. B., Aquilano, N. J. and Jacobs, F. R. *Production and Operations Management*. Boston, MA: Irwin McGraw-Hill, 1998.

Childe, S. J. *An Introduction to Computer Aided Production Management*. London: Chapman & Hall, 1997.

Duenyas, I. "Estimating Throughput Time of a Cyclic Assembly Station," *International Journal of Production Research*, 32 (1994), pp. 1403-1419.

Duenyas, I., and Keblis, M. F. "Release Policies for Assembly Systems," *IEE Transactions*, 27 (1995), pp. 507-518.

Frein, Y., DiMascolo, M., and Dallery, Y. "On the Design of Generalized Kanban Control Systems," *International Journal of Operations and Production Management*, 15 (1995), pp. 158-184.

Gstettner, S., and Kuhn, H. "Analysis of Production Control Systems Using CONWIP," *International Journal of Production Research*, 34 (1996), pp. 3253-3273.

Hopp, W. J., and Spearman, M. L. *Factory Physics: Foundations of Manufacturing Management*, Chicago: Irwin, 1996.

Im, J. H. "Lessons From Japanese Production Management," *Production and Inventory Management*, First Quarter (1989), pp. 25-29.

Karmarker, U. "Getting Control of Just in Time," *Harvard Business Review*, September/October (1989), pp. 122-133.

Keller, A. Z., and Kazzi, A. "Just In Time Manufacturing Systems: A Literature Review," *Industrial Management and Data Systems*, 93 (1993), pp. 2-32.

Monden, Y. *Toyota Production System: A Practical Approach to Production Management*, Atlanta, GA: Industrial Engineering and Management Press, 1983.

Ohno, T. *Toyota Production Systems: Beyond Large Scale Production*. Cambridge, MA: Productivity Press, 1988.

Ohno, K., Nakashima, K., and Kojima, M. "Optimal Numbers of Two Kinds of Kanbans in a JIT Production System," *International Journal of Production Research*, 33 (1995), pp. 1387-1401.

Philipoom, P. R., Rees, L. P., and Taylor, B. W. "Simultaneously Determining the Number of Kanbans, Container Sizes, and the Final Assembly Sequence of Products in a Just-In-Time Shop," *International Journal of Production Research*, 34 (1996), pp. 51-69.

Price W., and Gravel, M. "Modeling the Performance of a Kanban Assembly Shop," *International Journal of Production Research*, 33 (1995), pp. 1171-1177.

Rees, P. R., Philipoom, P. R., Taylor, B. W., and Huang, P. Y. "Dynamically Adjusting the Number of Kanbans in a Just-In-Time Production System Using Estimated Values of Leadtime," *IEE Transactions*, 19 (1997), pp. 199-207.

Roderick, L. M., Toland, J., and Rodriguez, F. P. "A Simulation Study of CONWIP versus MRP at Westinghouse," *Computers and Industrial Engineering*, 26 (1994), pp. 237-242.

Schniederjans, M.J. *Topics in Just In Time Management*. Boston: Allyn and Bacon, 1993.

Singh, N. *Systems Approach to Computer Integrated Design and Manufacturing*. New York: John Wiley & Sons, 1996.

Spearman, M. L., Woodruff, D. L., and Hopp, W. J. "CONWIP: A Pull Alternative to Kanban," *International Journal of Production Research*, 28 (1990), pp. 879-894.

Srinivasan, K., Kekre, S., and Mukhopadhyay, T. "Impact of Electronic Data Interchange Technology on JIT Shipments," *Management Science*, 40 (1994), pp. 1291-1304.

Stockton, D. J., and Lindley, R. J. "Implementing Kanbans Within High Variety/Low Volume Manufacturing Environments," *International Journal of Operations and Production Management*, 15 (1995), pp. 47-59.

Takahashi, K., Nakamura, N., and Izumi, M. "Concurrent Ordering in JIT Production Systems," *International Journal of Operations and Production Management*, 17 (1997), pp. 267-290.

Wantuck, K.A. *Just-In-Time: For America: A Common Sense Production Strategy*. Milwaukee, WI: The Forum, Ltd., 1989.

Yan, H. "The Optimal Number of Kanbans in a Manufacturing System with General Machine Breakdowns and Stochastic Demands," *International Journal of Operations and Production Management*, 9 (1995), pp. 89-103.

PART III

SELECTED JUST-IN-TIME TOPICS

5

Topics in JIT Supply-Chain Management: Part I

The purpose of this chapter is to describe the "front end" of JIT supply-chain management. A discussion of JIT supplier relationships is presented as well as a series of new strategies useful in the development of JIT supplier relationships.

WHAT IS SUPPLY-CHAIN MANAGEMENT?

With ever shortening product life cycles, complex corporate joint ventures, and stiffening requirements for customer service, it is necessary to consider the complete scope of supply-chain management, from supplier of raw material to in-store demand for product (see Davis, 1993; Lummus, Vokurka, and Alber, 1998). *Supply-chain* is a system whose parts include material suppliers, production facilities, distribution services, and customers linked together through the feed-forward flow of materials and the feedback flow of information (see Stevens, 1989). Managing these supply systems is called *supply-chain management*.

As Figure 5.1 depicts, a supply-chain can consist of multiple tiers of suppliers and vendors as well as multiple levels of distribution channels before the product reaches the end consumer. It is not uncommon to have up to 2,000 third-tier suppliers with in the framework of a single supply-chain.

As we move into an ever more competitive environment the management of the entire supply-chain becomes more critical. Issues surrounding the traditional goals of delivery, quality, speed, and flexibility take on a whole new level of complexity when dealing with a supply-chain. Traditionally lowering inventory levels, outsourcing materials or even changing production technology was consi-

FIGURE 5.1. Model of a Supply-chain

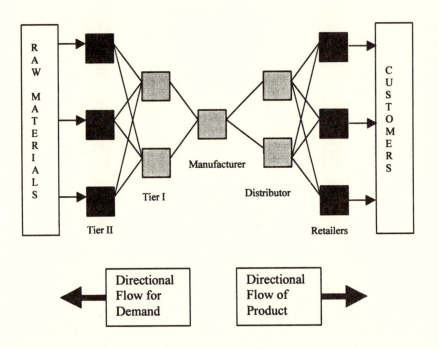

dered advantageous for most organizations. However, lowering the inventory level of one firm may raise the inventory level their supplier firm, which could make the move detrimental to the supply-chain. To make matters even more complicated, most organizations belong to several different supply-chains simultaneously.

WHY IS SUPPLY-CHAIN MANAGEMENT SO IMPORTANT?

Proper supply-chain management is critical if companies are to remain competitive in the future. The trend of increased competition and ever increasing customer expectations are going to force organizations to think about issues across their organizational boundaries and focus on those interfirm supply-chain issues. The competition of the future will be supply-chain to supply-chain, not firm to firm (see Schorr, 1998).

In recent years the topic of supply-chain management has received a great deal of attention. Research issues related to supply-chain management, such as forecasting, inventory control, and information sharing, are and have been researched for years. The difference between supply-chain management and traditional operations management lies in two dimensions: integration and coordination (see Lee and Ng, 1998). Integration is organizations working closely with both suppliers and customers to reduce or remove organizational boundaries. Coordination includes the flows within a supply-chain. The primary flows in a supply-chain are materials, finance, and information. The remaining portion of this chapter will focus on the flow of information and product between supply-chain entities.

JIT SUPPLIER RELATIONSHIPS/JIT PURCHASING

As one of the key elements of JIT, most companies now recognize the importance of developing long-lasting, productive relationships with their suppliers. There are at least eight criteria that determine an ideal supplier (see Holmstrom, 1998; Lummus, Vokurka, and Alber, 1998; Schorr, 1998, pp. 29-43):

1. *Delivery.* Are firms able to deliver on the date they promised? In a JIT environment this is crucial to the successful operation of the entire system. Suppliers need to be able to make frequent deliveries and be flexible enough to cover unexpected demand surges.

2. *Quality and reliability.* Are suppliers are able to deliver the correct amount every time and are the shipments delivered of high quality? In JIT we strive for accuracy and reliability of products being shipped to our organization. Incoming products should neither have to be inspected nor counted for accuracy. Often overlooked in this equation is the quality of supplier customer service. Even the best planned systems are prone to errors, and suppliers must be able to respond to those problems and correct problem situations.

3. *Price.* Are the prices of the supplier fair? No longer is the focus of our purchasing managers to drive the price as low as possible. "Fair" price refers to the situation when the supplier can make a profit, but your organization can remain competitive in the marketplace. This creates a win-win situation for the downstream organization as well as their suppliers. Organizations are expected to work together to reduce the costs over the long haul.

4. *Leadtime.* What are the lead-time requirement situations of the suppliers? JIT suppliers should share information regarding their lead-time to their customers. In today's competitive environment customers place special orders on a frequent basis. In order to respond to these special orders, a JIT organization must have a good understanding of their supplier's lead-time for the proper procurement of material and supplies.

5. *Location.* What is the proximity of the suppliers? To maintain frequent, small deliveries the location of suppliers is a major concern for JIT organizations. A JIT organization must not only be concerned with its immediate suppliers but with their suppliers as well, to maintain a high performing supply-chain.

6. *Technological capabilities.* What are the technical capabilities of the suppliers? Are the suppliers able to keep up with demand patterns and are they able to adjust to new demand patterns. In other words, do your suppliers have the technical capabilities to handle rapid changing market conditions.

7. *Financial stability.* What is the likelihood that suppliers will be out of business in the near future? Even if a supplier is not entering bankruptcy, if they are on the verge they wont be able to provide the quality and reliability expected for the organization to be a world class competitor.

8. *Supply-chain management.* Are the suppliers willing to participate in improvement efforts for the entire supply-chain? Good suppliers are willing to share information

to all members of the supply-chain. This ensures information can be coordinated among all members of the supply-chain.

These are merely guidelines for developing strong supplier relationships. The guidelines mentioned above closely resemble the characteristics of a good JIT purchasing relationship.

The concept of *JIT purchasing* is not new; in fact we can trace evidence of its use in the United States industry back to the late 1970s. The concept of JIT purchasing is one where the supplier and purchaser develop a working relationship that aggressively attempts to make both parties prosperous (Schniederjans, 1993). Some of the critical characteristics of this relationship include: long-term contracts, improved accuracy of order filling, improved quality, ordering flexibility, small lots ordered frequently, and continuous improvement in the partnership.

The combination these elements are what comprises a good JIT purchasing relationship. Both the fundamentals of JIT purchasing and above mentioned characteristics of a good supplier point toward establishing solid relationships with suppliers. If competition in the future will require the capabilities of many firms in the supply-chain, it will be beneficial for firms to develop these partnering relationships. Organizations must realize that we can cooperate with our suppliers and sometimes competitors, but still remain competitive. The emphasis is to create a win-win situation with suppliers.

JUSTIFICATION FOR JIT PURCHASING

JIT purchasing, if used correctly, can be a component that will help create superior firm performance. Germain and Droge (1998) examined the, organizational design and performance of JIT buying and non-JIT buying firms. They suggest that JIT purchasing is appropriate across a wide spectrum of competitive environments. They also suggest a firm can still receive benefits even if their suppliers are not JIT based.

Further comparisons suggest the organizational design of JIT buying firms is significantly different from non-JIT buying firms. In JIT buying firms the level of performance control is very high. JIT buying firms are more likely to monitor profitability, costs, and productivity level, from both internal data, and that relative to the competitors. They also tend to be more in tune with supplier performance keeping a close eye on supplier price, delivery, and manufacturing capacity. In addition, JIT buying firms are more likely to be proactive in developing a purchasing mission and establishing a performance monitoring systems.

Finally, the performance is higher for JIT buying firms versus not JIT buying firms. Fisher and Raman (1996) discovered that JIT buying firms carry less inbound inventory and managers assess their long-range overall performance to be higher than non-JIT buying firms. The evidence from the study indicated that JIT buying firms reported superior performance in terms of both market share and financial performance measures.

The results of this research have far reaching effects on how we view JIT purchasing in the supply-chain. First and foremost, we can clearly see that overall JIT buying firms achieve superior performance relative to their non-JIT buying counterparts. Not only do they have higher performance that non-JIT buying firms, it does not seem to make a difference in what situation we use JIT buying as our purchasing strategy. Interestingly though, JIT buying firms are more proactive in monitoring the performance levels in their organization as well as supplier organizations. To monitor this performance, the JIT buying firms have developed and standardized good metrics of performance. All of this suggests that the JIT buying firms do a better job of recording, tracking, and keeping up to date information on the performance of their firm relative to their competitors.

NEW STRATEGIES

As JIT purchasing began to pick up momentum and its popularity grew, some problems arose that reduced its overall success. Some of the related issues include: lack of support from suppliers, product quality, lack of communication, and transportation and logistics problems (see Godar and Stamm, 1991). As a result researchers began to discover that even though JIT purchasing can be successful in a variety of firm contexts, some related issues needed to be addressed to ensure a JIT purchasing program has success. Early attempts at addressing the problems included developing long-term contracts with suppliers, supplier training and certification, and synchronizing deliveries with production.

A whole new set of issues has risen as the importance of the overall supply-chain grows. As mentioned earlier we will soon be competing supply-chain to supply-chain, not company to company, so what was initially effective as a method for improving individual firm performance may not be beneficial for the overall supply-chain. As a result, some of the initial JIT purchasing strategies may have lost some of their appeal because we are dealing with several levels of a supply-chain. The remaining portion of this section highlights the most current topics in JIT purchasing that help handle this multifirm, multinational problem.

CONTROLLING VARIABILITY ACROSS THE SUPPLY-CHAIN

In an effort to compete, organizations are forced to deal with issues that span current organizational boundaries. Practitioners and researcher alike are beginning to recognize the impact of variability on performance across different levels of the supply-chain. The "bullwhip effect," "clockspeed amplification," and methods to handle these supply-chain dynamics are discussed in this section.

The Bullwhip Effect

The *bullwhip effect* refers to situations when the ordering patterns at the retail level or the manufacturing level may be very stable but upstream in the supply-chain the order variability is greatly amplified (see Lee, Padmanabhan, and Whang, 1997a, 1997b; Metters, 1997; Towill, 1997a, 1997b; Wilding, 1998). Figure 5.2 provides a graphical illustration of the bullwhip effect. The ordering patterns show a common theme: the variability in the ordering process at the upstream sites is always greater than the downstream sites.

To visualize the bullwhip effect, consider a simple example of consumers deciding to switch to a new computer because of a faster microprocessor. At the consumer end of the supply-chain this change of preferences creates a ripple in the supply-chain (i.e., orders for the new compute rise). Due to information lags, delivery lags, miscalculation of the size of spike in demand and other factors the ripple gets magnified. The effects accumulate, resulting in large swings in orders further back in the supply-chain. A example from the machine tool industry can help see the bullwhip effect in action. From 1961 to 1991 the gross domestic product had swings in the range of 2 to 3 percent, the automotive industry had swings of 20 percent, and the machine tool industry had swings in orders of 60 to 80 percent (see Fine, 1998). As a result this ripple effect forced many organizations out of business.

Such an effect makes JIT purchasing efforts more difficult upstream in the supply-chain. JIT purchasing requires ordering flexibility, improved accuracy in order filling, and frequent ordering of smaller lots to be successful. If the variability upstream in the supply-chain in the ordering process is extreme, satisfying these conditions may be near impossible. Situations may exist where the organization may need large amounts of finished goods inventory to satisfy those requirements. The major causes of the bullwhip effect have been suggested by Metters (1997) and Lee, Padmanabhan, and Whang, (1997a, 1997b) to include the following:

1. *Forecast updating.* Forecasts are based on the order history from the company's immediate customers. However there is a time lag between production and the processing of the orders at the upstream site in the supply-chain. This time lag forces downstream sites to order sooner in quantities greater than what is needed for production (Fisher and Raman, 1996). The processing of these demand signals is a major contributor to the bullwhip effect.
2. *Quantity ordering.* Due to costs to process orders may companies have a periodic ordering policy. In such situations a supplier can face a highly erratic stream of orders from a customer. If there are demand spikes, the variability increases and forces the organization to adjust their periodic order amount, this adjustment adds to the bullwhip effect.
3. *Forward buying and promotions.* Manufacturers frequently offer promotions and retailers quite frequently forward purchase seasonal items. This results in customers buying in quantities that do not reflect their immediate needs. Usually they end up buying in large quantities and stock up for the future. This results in customer's buying pattern not reflecting the consumption pattern, the variation of purchased quantities is larger than the variation of the consumption rate creating the bullwhip effect.

FIGURE 5.2. Bullwhip Effect in Supply-Chains

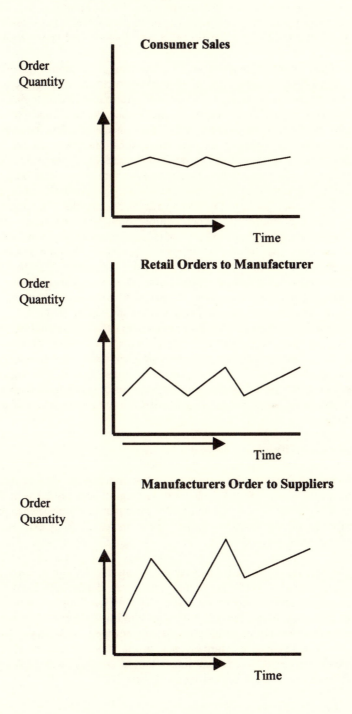

As managers gain greater understanding of the bullwhip effect, better strategies to cope with this phenomenon will be developed. As presented in Table 5.1, various control mechanisms are available to help counteract the effects of the bullwhip effect, including information sharing, logistics, and operational strategies.

In a sense the whole purpose of JIT purchasing is to control the effects of the bullwhip effect. By avoiding the multiple demand forecasting, we are suggesting the organizations readily employ new information technologies such as EDI, CAO and Internet technologies to ensure data is shared and coordinated throughout the entire supply-chain. While the concept of using these information technologies is not new to JIT purchasing situations, the excessive cost associated with these technologies has made it impractical to implement supply-chain wide. The Internet has now made these technologies accessible to every member of the supply-chain. This rapid transmission of demand data is going to be necessary for supply-chains to compete against one another in the future.

The importance of information coordination throughout the supply-chain is critical to achieving superior supply-chain performance (see Towill, 1997a, 1997b). The most significant performance is gained by making available the undistorted and up-to-date demand data at every echelon within the supply-chain. To achieve an effective supply-chain, clean information flow must be guaranteed. This requires a sense of openness and commonality of vision, which should be shared throughout every level of the supply-chain.

What are the implications for the supply-chain based on the bullwhip effect? First, every company is dependent on others in a supply-chain. As a result our JIT purchasing strategies are meaningless unless we apply them to multiple levels of the supply-chain. Second, organizations must look at their capabilities supply-chain wide. To utilize the entire capabilities of the supply-chain, organizations must be willing to cooperate and assist the weaker members of the supply-chain.

Clockspeed Amplification

Clockspeed refers to the rate at which an industry evolves (Fine, 1998; Mendelson and Pillai, 1999). Clockspeeds are significantly faster near the end customer in the supply-chain. Quite simply, consumer preferences change much quicker for consumable items as opposed to raw materials. In an extreme example, clothing retailers face a very rapidly changing environment. Consumer preferences in the fashion industry change very rapidly, forcing retailers to change product variety frequently. Upstream from the clothing retailers are the textile manufacturers. While consumer preferences change rapidly, the manufacturers process and models do not change as rapidly. Suppliers to the textile mills are the raw material providers of wool and other fabrics. Obviously the production of wool and other fabrics has not changed substantially in many years. The end result is the further back in the supply-chain the slower the clockspeed of products. A graphical illustration of *clockspeed amplification* for the clothing example above is illustrated in Figure 5.3.

TABLE 5.1. Controlling Mechanisms for the Bullwhip Effect

Information Sharing Strategies
    Point of Sale Data Transfer
    Electronic Data Interchange
    Internet/Extranet Data Exchange
Logistics Strategies
    Vendor Managed Inventory
    Direct Purchasing
    Logistics Outsourcing
Operational Strategies
    Lead Time Reduction
    Frequent Deliveries
    Everyday Low Pricing
    General JIT Improvement Efforts

The question arises then is what is the effect of clockspeed amplification on JIT purchasing practices. To a large extent the question being asked is how do organizations in JIT purchasing environments deal with rapid product changes. A good JIT organization has flexibility incorporated into its operations to handle changes in consumer preferences and expectations. However, flexibility by itself does not ensure the supply-chain will be able to handle the change, improved chain cooperation is also needed. An organization can still remain competitive while cooperating with suppliers, vendors, and sometimes competitors. The increased sharing of information and technology can and will lead to improved supply-chain performance. These variability issues are discussed in greater detail in Chapter 6.

INTERNATIONAL JIT PURCHASING

It is no secret that we are no longer competing with the firm across the street, but rather on a global basis. As a result, organizations are faced with an entirely new set of challenges to remain competitive. While the principles of JIT purchasing can be applied in a variety of organizational contexts and situations, multinational firms often have suppliers and facilities in underdeveloped, third-world nations that have limited capabilities.

Multinational firms are beginning to use more third-world suppliers in the production of their goods and services. Suppliers located in these underdeveloped countries are very often rich in raw material resources and are able to provide labor at extremely low prices making their use in the supply-chain very attractive. U.S. manufactures have continued to use more third-world labor and materials as competition and customer expectations increase. Many of these

FIGURE 5.3. Clockspeed Amplification in Supply-Chains

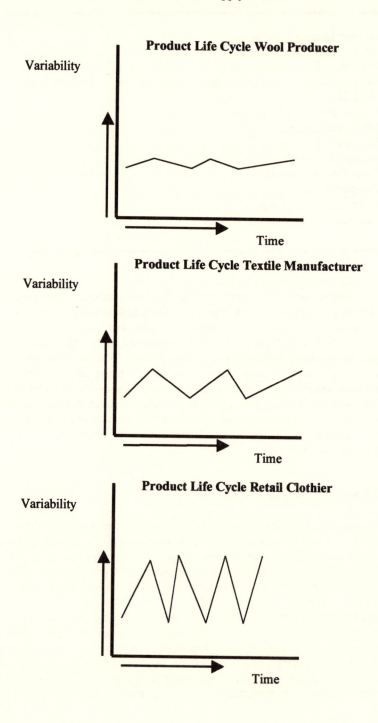

multinational corporations are moving substantial pieces of their production systems to these underdeveloped countries to capitalize on the cheap labor. Some of the larger industries that lead the way toward this trend include the automotive, textile, and computer industries.

Unfortunately, these third-world countries often have very poor infrastructure, poor quality, and very unreliable transportation and communication systems. In fact, in the early 1990's in a study of JIT practices in Mexico, Vargas and Johnson (1993) report 82 percent of all companies in Mexico maintain large raw stocks of all products. This indicates that the organizations in Mexico are struggling to transition to a JIT environment.

Companies that have moved their operations to Mexico and other third-world countries are finding that the supplier bases are typically very weak and unreliable (see Lawrence and Lewis, 1996). These suppliers cannot provide the level of quality and reliability required of a traditional manufacturing environment, let alone the stringent requirements of a JIT purchasing environment. This often forces these organizations to use more expensive domestic and foreign suppliers to provide shipments to enable a JIT purchasing environment. The long lead-times and unreliable infrastructure of the third-world countries make timely delivery of inventory and information near impossible. All of these counter the potential savings acquired by the cheap labor and capacity making the move ineffective.

With all of these problems how can an organization effectively use third-world suppliers and labor as an effective means for JIT purchasing? A recent study by Lawrence and Lewis (1996) highlights several key issues related to JIT purchasing in third-world countries.

1. *JIT deliveries from suppliers.* Most third-world suppliers (the study highlights Mexican suppliers) are unable to meet the needs most of the basic needs of corporations competing in world markets. All production firms in Mexico have materials delivered from international sources and many need more than 75 percent of their requirements shipped from international suppliers. These suppliers were unable to deliver the basic production requirements, much less deliver product in a JIT manner. However, the aggressive users of JIT have been able to achieve some degree of success in these undeveloped nations. They devote a significant amount of resources and time towards developing the local suppliers materials management systems and total quality management program. The majority of the firms maintained at least one week of safety stock because of the quality and reliability problems of the local suppliers. In combination with transition times these firms can often have up to three weeks of WIP in the pipeline.

2. *Supplier involvement with improvement activities.* Most third-world suppliers are unable to contribute to the continuous improvement activities in the supply-chain because of the lack of resources or technical capabilities. Not only do suppliers have little involvement in improvement activities, they tend not to be active in developing new products and better designs for existing products.

In a comparison with most U.S. plants the cost to develop a solid JIT purchasing program is substantially more in Mexico and other lesser-developed countries. Despite this limitation, most facilities have lower total costs than their U.S. counterparts, primarily due to the extremely low labor rates. However, these facilities generally had poor supplier reliability (80 percent of deliveries were reported late), longer lead-times than U.S. counterparts and kept large safety stocks because of poor delivery reliability.

Other considerations to account for when using third-world suppliers are the legal system of the developing nations, language barrier, and computer literacy. Often times the legal system makes the use of a paperless system infeasible. The inability to use modern technologies such as EDI and e-commerce reduces the accuracy of orders and increases the total cycle time of the system. The language barrier and poor computer literacy of the developing nations is another problem often encountered by firms using third-world suppliers. As Garg, Kaul, and Deshmulch, (1998) and Schniederjans, (1998, pp. 71-85) have described, the barriers between U.S. and third-world suppliers can be significant (see Table 5.2).

The stringent requirements of a JIT purchasing program have made its use extremely difficult in developing nations such as Mexico and India. For a JIT purchasing program to be effective in third-world nations a substantial amount of time an effort is needed to develop both the infrastructure of the firm as well as the country. JIT purchasing firms are forced to allocate a substantial amount of time and capital towards the development of the suppliers capabilities, so they can conform to the high standards set by JIT firms.

TABLE 5.2. Comparison of Developed and Third-World Suppliers

Type of Measure	Developed Countries	Third-world
Literacy Level	Very High	Very Low
Language	One Language	Many (regional & local)
Level of Advanced Technology	Very High	Moderate to Low
Computer Literacy	Very High	Fair
Level of Infrastructure	Excellent (excellent transportation and communication)	Fair (Still in developing stage)
Facilities (plant, equipment, etc.)	Very Advanced	Poor to Good
Product Quality	High and Reliable	Unreliable
Delivery Reliability	Good to Excellent	Poor to Good

SUPPLIER STRATEGIES

The literature abounds with a variety of supplier strategies that can be employed to improve operations. This sections describes two of the more currently used strategies: supplier incentive/penalty systems and vendor managed inventory.

Supplier Incentive Systems

We have demonstrated the importance of supplier relationships as a component for JIT purchasing. Often supplier incentives are used to in the formation of a buyer/seller contract to ensure the supplier is able to meet production quotas, on-time delivery requirements and quality requirements.

On-time performance is recognized as one of the critical components of JIT purchasing. One of the reasons that JIT purchasing has been difficult to implement is that timing is critical to the buyers operations, however the supplier controls the timeliness of deliveries. Since little or no excess inventory is kept to compensate for supply uncertainty, late deliveries cause work stoppages and other problems that affect the performance of the supply-chain. To compensate for handling this timeliness issue, organizations create buyer/seller relationships that will ensure on-time deliveries (see Grout, 1996). As a method of contractual control many organizations use a *supplier incentive/penalty systems* to create the desired response from the supplier. In the early 1990s nearly 50 percent of all JIT purchasing contracts use penalties for nonconformance as an incentive for on-time delivery (Freeland, 1991).

An examination of this supplier-based incentive/penalty system yields insights into the timeliness problem faced by buyers. Once buyers reduce raw material inventories to very low levels, the buyers then depend heavily on the timeliness of suppliers to maintain an efficient and effective production system. In most instances the buyers may attempt to achieve 100 percent on-time delivery to satisfy their production requirements. Under most circumstances this is an unreasonable assumption, unless the suppliers of the goods are provided an incentive clause in their contract and a their production system is capable of uniform production.

Most production systems are not capable of production a uniform rate and in these instances 100% on-time delivery is generally not possible nor is it optimal for the system. Under these conditions the buyer should select desired probabilities of on time performance and an appropriate incentive schemes. Once the incentive clause is chosen the supplier responds by selecting a production rate that will minimize the suppliers expected costs. For a complete model of this type of system see Grout (1996).

This lends some insights into the problem of manufactures placing unrealistic assumptions on suppliers. Under most conditions suppliers are unable to provide 100% on-time delivery for their customers. However, traditionally manufacturers have placed these unrealistic assumptions on their suppliers and as a consequence product quality suffers and poor supplier/customer relationships are created. A mutual understanding between buyers and sellers creates an environment that can lead to superior supply-chain performance. The negotiation process between manufactures surrounding delivery requirements and contract incentives creates a sense of trust and understanding required for superior supply-chain performance. As this relationship grows the organizations can pool resources to improve on-time delivery performance and as a result contractual incentives/penalties can be adjusted accordingly. For organizations using JIT purchasing it is this integration between buyer and supplier that will lead to superior supply-chain performance (see Ha and Kim, 1997).

Vendor-Managed Inventory

Realizing that competition is forcing companies to shorten lead-times, improve product quality and variety, and increase the level of customer service, organizations are establishing better channels between supply-chain members. These links can consist of information sharing or involve sharing processes or activities. Creating these links can be a source of competitive advantage for most firms (see Porter and Millar, 1985). The use of information technologies is one of the primary methods in which these links can be established.

Vendor-managed inventory (VMI) (often referred to as *continuous replenishment*) is a method for controlling the flow of information between supply-chain members (see Cachon and Fisher, 1997; Clark and Hammond, 1997; Holmstrom, 1998). The VMI process is essentially a two-phased process: Retailers report daily to the supplier their recent demand and current inventory position, and suppliers use these data to determine shipment replenishments to each retailer of retailers' warehouse. This is the perfect complement to an already established JIT purchasing program. Both retailers and manufacturers have up-to-date demand information that allows them to effectively plan their production flow and timing of shipments.

This system varies significantly from traditional EDI systems in that it requires substantial changes in the way data are transmitted or analyzed in the supply channel. Under traditional EDI systems order quantities are determined by methods that were in place before the adoption of EDI. This indicates why in most situations organizations indicate little or no impact from traditional EDI systems on company performance. The rate of data exchange and accuracy of demand data are the principal differences between traditional EDI systems and VMI systems.

Since its inception, VMI has been adopted in many industries, including the grocery industry and textile industries. How do firms go about establishing a VMI system? Campbell's VMI program has four critical components (see Cachon and Fisher, 1997):

1. *Every Day Low Pricing.* Products are sold to retailers on and *every day low pricing* (EDLP). As mentioned earlier EDLP is a way for retailers and suppliers to control demand amplification in the supply-chain. This makes demand forecasting much easier throughout the supply-chain. However, it should be mentioned the EDLP does not restrict retailers from offering promotions.
2. *Data transmission.* Every weekday morning retailers send through an EDI system, their current inventory position and demand data for their distribution centers. This

gives Campbell's the opportunity to adjust their production schedule so they can maintain a low inventory position and still ship the product JIT to avoid stockouts.

3. *Shipment resupply.* Campbell's sends the resupply shipment to the distribution center or directly to the retailer based on the up to date demand information and inventory status.

4. *Optimize truckloads.* Only full truckloads are shipped to either the distribution center or retailer. This is put in place to maintain low transportation costs.

Notice that a component of the VMI system at Campbell's is the use of an EDI system. While it was indicated that EDI in itself had little or no impact on firm performance, the use of EDI while reengineering the supply channel proves to be very effective. To ensure the most up-to-date demand data are distributed throughout the supply-chain, data are transmitted at *point of sale* (POS) and not from traditional methods. Since Campbell's introduction of its continuous replenishment program, it's had an estimated 50 percent reduction in retailers' inventory, an increase in inventory turns by 13, and stockouts decreasing by 2.1 percent.

While VMI has been introduced at the retail level it has not yet been introduced at all levels of the supply-chain. The up-to-date demand data transmission has the potential to receive extremely large benefits when implemented across the supply-chain.

WHERE ARE WE HEADED?

As preceding discussion points out, the traditional view of competition is changing and the ramifications on JIT practices are far reaching. The relationship organizations create with their suppliers will be one of the elements in creating a competitive advantage.

The days of being able to provide a single product with little variety are long since gone. To survive in the marketplace today organizations are forced to provide substantial product variety and must be able to introduce products very rapidly. Product life cycles for many products can be compared to fruit flies; like the fruit fly they are introduced to the market and soon die with another generation of evolved products taking their place (see Fine, 1998). This phenomenon forces organizations and their suppliers to be able to react quickly to changing market conditions. To be able to respond to the changing conditions effective supplier relationships are necessary.

JIT purchasing programs have enabled organizations to establish these strong supplier relationships. The issues surrounding an effective JIT purchasing program remain much the same. Organizations still need to establish long-term contracts, improved accuracy of order filling, improved quality, ordering flexibility, small lots ordered frequently, and continuous improvement in the partnership to be successful. However, these issues are made much more complex, extending them internationally and to multiple echelons in a supply-chain. In the next decade the organizations in their respective supply-chains will need to address the issues surrounding the globalization of the supply-chain and advanced data communication methods.

Globalization of the Supply-chain

In an effort to reduce prices to meet consumer demand, organizations are using an ever increasing amount of third-world suppliers for raw materials and subassembled products. The extremely low labor rates associated with third-world suppliers make them a very attractive alternative to domestic suppliers. However, their product quality is generally not as good and the lead-times associated with their products are usually substantially longer than regional suppliers. As demonstrated earlier, JIT purchasing can still be very effective for these underdeveloped countries, but has reduced overall effectiveness in JIT purchasing. The poor infrastructure, in particular the communication networks makes it virtually impossible for competing supply-chains to provide up-to-date demand information in order to be competitive.

To gain the benefits associated with using third-world suppliers, organizations will need to make substantial investments in both the national infrastructure (e.g., roads, communication networks, etc.) and organizational infrastructure (e.g., material management systems, modernizing equipment, etc.). The full benefits of JIT purchasing and JIT production are received when the inefficiencies associated with poor capabilities are eliminated. As costs and labor rise in these foreign countries, the substantial benefits organizations are currently receiving will diminish substantially, if not be totally eliminated. Since many of the third-world nations are rich in raw materials and resources, not ordinarily found domestically, it is advantageous to invest in their infrastructure. Savings associated with increases in productivity and efficiency will offset the costs associated with the investments in infrastructure.

Advanced Data Communication Methods

Matching supply with demand is becoming more difficult as new product introductions and modifications to existing products accelerate (see Fisher et al., 1997). In addition to clockspeed amplification, the bullwhip effect has created a need for advanced methods of sharing data throughout the entire supply-chain. Initial solutions have included using technologies such as EDI, VMI, and POS data in forecasts. In particular, vendor managed inventory is a very advanced method for reducing the overall level of inventory in the supply-chain, while maintaining a high level of service.

Currently, the capabilities of advanced information technology in JIT have not yet fully explored. VMI is used fairly exclusively in the grocery and clothing industries. As the benefits of VMI are realized, more industries will begin to use VMI in their supply-chain. The effectiveness is predicated on the willingness of managers to share demand, product, and inventory information.

The evolution of VMI will be the use of *extranets* (i.e., secured Intranets) to exchange not only demand information, but customer service and product information as well. Many large organizations, such as Ford are beginning to use Extranets as a method to share demand information, customer service records, and inventory data (see "Ford Turns to Extranets," 1998). Issues surrounding extranet use include: building a secure system and system unreliability. However, the cost to implement is very reasonable when compared to traditional EDI systems.

SUMMARY

This chapter presented an overview of JIT supplier practices as they relate to supply-chain management. Some current topics and techniques were also discussed.

The issues surrounding the development of good supplier relationships still remain, but have become much more complex in today's marketplace. In the early years of JIT implementation almost any change to the existing system had dramatic effects because the concept of waste elimination was foreign to most companies. As the marketplace has become more competitive the firms and now supply-chains are looking for new ways to stay ahead of the competition. Organizations have begun to look at all members of the supply-chain collectively as method to remain competitive.

This chapter has given an overview of the complexities surrounding the development of effective JIT supplier relationships in a supply-chain. Clearly issues such as third-world suppliers and globalization of the supply-chain have made the use of JIT purchasing much more difficult than is years past. The purpose of this chapter was to present some of the new issues surrounding JIT purchasing and present some techniques to help alleviate the situation. In the next chapter, we continue the discussion of additional supply-chain issues.

REFERENCES

Blackburn, J. The Quick Response Movement in the Apparel Industry: A Case Study in Time-Compressing Supply-chains. In Blackburn, J. *Time Based Competition: The Next Battleground in American Manufacturing.* (pp. 123-134) Homewood, IL: Irwin, 1991.

Cachon, G., and Fisher, M. "Campbell Soup's Continuous Replenishment Program: Evaluation and Enhanced Decision Rules," *Productions and Operations Management*, 6 (1997), pp. 266-276.

Clark, T. H., and Hammond, J. H. "Reengineering Channel Ordering Process to Improve Total Supply-chain Performance," *Productions and Operations Management*, 6 (1997), pp. 248-265.

Davis, T. "Effective Supply-chain Management," *Sloan Management Review*, 82 (1993), pp. 35-46.

Fine, C. *Clockspeed: Winning Industry Control in the Age of Temporary Advantage.* Reading, MA: Perseus Books, 1998.

Fisher, M., Hammond, J., Obermeyer, W., and Raman, A. "Configuring a Supply-chain to Reduce Demand Uncertainty," *Productions and Operations Management*, 6 (1997), pp. 211-225.

Fisher, M., and Raman, A. "Reducing the Cost of Demand Uncertainty through the Accurate Response to Early Sales," *Operations Research*, 44 (1996), pp. 89-99.

"Ford Turns to Extranets." *Informationweek*, (August 10, 1998), pp. 30.

Freeland, J. R. "A Survey of Just-In-Time Purchasing Practices in the United States," *Production and Inventory Management Journal*, 32 (1991), pp. 43-49.

Garg, D., Kaul, O. N., and Deshmukh, S. G. "JIT Implementation: A Case Study," *Production and Inventory Management Journal*, 39 (1998), pp. 26-31.

Germain, R., and Droge, C. "The Context, Organizational Design, and Perform-ance of JIT Buying versus Non-JIT Buying Firms," *International Journal of Purchasing and Materials Management*, 34 (1998), pp. 12-18.

Godar, D., and Stamm, C. "The Just-In-Time Philosophy: A Literature Review," *International Journal of Production Research*, 29 (1991), pp. 657-676.

Grout, J.R, "A Model of Incentive Contracts for Just-In-Time Delivery," *European Journal of Operations Research*, 96 (1996), pp. 139-147.

Ha, D., and Kim, S. L. "Implementation of a JIT Purchasing: An Integrated Ap-proach," *Production Planning and Control*, 8 (1997), pp. 152-157.

Holmstrom, J. "Implementing Vendor-Managed Inventory the Efficient Way: A Case Study of Partnership in the Supply-chain," Production and Inventory Management Journal, 39 (1998), pp. 1-5.

Lawrence, J. L., and Lewis, H. S. "Understanding the Use of Just-In-Time Pur-chasing in a Developing Country," *International Journal of Operations & Production Management*, 16 (1996), pp. 68-90.

Lee, H. L., and Ng, S. M. "Preface to Global Supply-chain Management," *POMS Series in Technology and Operations Management*, 1 (1998), pp. 45-56.

Lee, H. L., Padmanabhan, V., and Whang, S. "The Bullwhip Effect in Supply-chains," *Sloan Management Review*, 86 (1997a), pp. 93-102.

Lee, H. L., Padmanabhan, V., and Whang, S. "Information Distortion in a Sup-ply-chain: The Bullwhip Effect," *Management Science*, 43 (1997b), pp. 546-558.

Lummus, R. R., Vokurka, R. J., and Alber, K. L. "Strategic Supply-chain Plan-ning," *Production and Inventory Management Journal*, 39 (1998), pp. 49-58.

Mendelson, H., and Pillai, R. "Industry Clockspeed: Measurement and Opera-tional Implications," *Manufacturing and Service Operations Management*, (in press).

Metters, R. "Quantifying the Bullwhip Effect in Supply-chains," *Journal of Op-erations Management*, 15 (1997), pp. 89-100.

Porter, M. E., and Millar, V. E., "How Information Gives You Competitive Advantage," *Harvard Business Review*, 63 (1985), pp. 149-160.

Schniederjans, M. J. *Topics in Just-In-Time Management*. Boston: Allyn and Bacon, 1993.

Schorr, J. E. *Purchasing in the 21st Century*. New York: John Wiley & Sons, 1998.

Stevens, J., "Integrating the Supply-chain," *International Journal of Physical Distribution and Materials Management,* 19 (1989), pp. 3-8.

Towill, D. R. "FORRIDGE - Principles of Good Practice in Material Flow," *Production Planning and Control,* 8 (1997a), pp. 622-632.

Towill, D. R. "The Seamless Supply-chain - The Predator's Strategic Advantage," *International Journal of Technology Management, Special Issue on Strategic Cost Management,* 13 (1997b), pp. 37-56.

Vargas, G. A., and Johnson, T. W. "An Analysis of Operational Experiences in the US/Mexico Production Sharing (maquiladora) Program," *Journal of Operations Management,* 11 (1993), pp. 17-34.

Wilding, R. "Chaos, Complexity and Supply-chains," *Logistics Focus,* 6 (1998), pp. 8-10.

6

TOPICS IN JIT SUPPLY-CHAIN MANAGEMENT: PART II

The purpose of this chapter is to describe the design and reengineering of the JIT supply-chain. A discussion of supply-chain design and reengineering issues and methodologies are presented.

THE "IDEAL" SUPPLY-CHAIN

What is the ideal supply-chain? Every organization is presented with a set of issues that are critical to the design of the perfect supply-chain. There is no one design that will ensure business success. However, supply-chain designers must observe the old adage that the chain is only as strong as its weakest link. Building a company or capability without regard for the entire supply-chain will spell disaster. In fact, it can be argued that there is no issue more critical to organization success than the design of an effective supply-chain from final consumers all the way back to the providers of raw materials (see Fine, 1998; Handfield and Nichols, 1999). The ideal supply-chain then takes into account all the relevant issues related to material suppliers, production facilities, and distribution systems for all entities in the supply-chain.

Building a superior supply-chain can lead to a substantial competitive advantage. We can see an example of this in how the Dell computer company has used their supply-chain design along with JIT to achieve a substantial competitive advantage (see Fine, 1998; Margretta, 1998).

Dell in a very short period of time, has been able to make a small start up business grow into a $12 billion business. From the period of 1990 to 1998 its stock price had increased over 26,000 percent higher than that of all the major competitors in the market. Most amazingly Dell is able to customize orders and still is able to provide large shipments to their customers on time. How is Dell able to achieve this remarkable success? The answer lies in superior supply-chain design and the use of JIT as the driving force.

Dell uses a strategy of customer focus, supplier partnerships, mass customization, and JIT in a unique manner that has created their computer empire. The model in which Dell uses is not extremely complex, but it is an entirely new way of conducting business in the computer industry. Since the inception of the computer industry, the competitors have been extremely vertically integrated. Since the technology and components were new, the manufacturers built large vertically integrated enterprises that built everything from the computer housings, memory chips, and disk drives. The traditional computer industry supply-chain is depicted in Figure 6.1 (see Bendiner, 1998; Margretta, 1998; Phillips, 1998).

The Dell supply-chain is considered to be a direct model (see Figure 6.2) of a supply-chain. Rather than being highly vertically integrated, Dell focuses on activities in which it can add additional value to the customer, above and beyond the production of computers. This leveraged relationship between suppliers and customers is what Dell has termed its virtual organization. Virtual integration means piecing together a business with customers and suppliers and treating them as if they are part of your organization.

There are many components that constitute the successful reign and implementation of a supply-chain. Dell has pieced together its supply-chain by developing a new model for conducting business in the computer industry. Dell has gained its competitive advantage by using the following elements: the direct model, advanced customer service methods, strong supplier relationships and advanced demand forecasting (see Fox, 1998; Margretta, 1998). Each of these components of a successful supply-chain operation will be discussed in turn.

Direct Model

Rather than build an elaborate system of retailers, distributors and wholesalers Dell uses the *direct model* to market and sell its products directly to customers (i.e., no wholesalers). Individuals and organizations use either a phone call or the Internet to access Dell directly to place orders. The finished goods are assembled within the supply-chain and are shipped directly to the customer. In this manner customers are able to get their materials very rapidly and at a very reasonable cost. Purchase price is kept to a minimum because distribution system costs are essentially removed from the system. Customers are also able to get customized models rather than display room standard models because all orders are make to order. Since Dell carries very little finished goods inventory, as consumer preferences change they are able quickly to change models and meet consumer needs. This enables Dell to capitalize fully on their JIT production environment. The lower inventories required for JIT imply they do not

FIGURE 6.1. Traditional Computer Supply-Chain

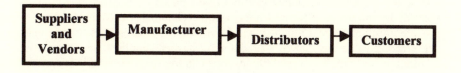

FIGURE 6.2. Dell's Direct Model of a Supply-Chain

have to sell the old antiquated models at reduced rates, they simply begin immediate production of the new models.

Company Growth

Dell has been able to grow at an extreme rate since the very inception of the
company. Why? The answer lies within how it has structured their company. It
has less people and facilities to manage than its vertically integrated competitors. Dell has developed JIT-type partnering relationships with its supplier of
component parts that enables it to get the highest quality component parts for its
products. Rather than having to build a new facility every time a component
part becomes outdated it uses its supplier network to make sure it always has the
most current and highest quality parts.

Customer Service

This is a unique component of Dell model that helps with the creation of
their vertical model. Dell indirectly employs thousands of customer service
representatives, but only a few work directly for its computer division. This
keeps the overhead down and allows Dell to focus on those value added services
they provide the customers.

Supplier Relationships

Generally firms outsource their problems or difficult components to regional and specialty suppliers. Dell makes agreements with its suppliers to pro-

vide a percentage of its production requirements every year, even when demand exceeds supply. In turn Dell provides its supplier with its daily production requirements reducing the total amount of inventory in the system and also reducing the impact of the bullwhip effect (see Chapter 5). All of this is made possible by the use of advanced information technologies that enables Dell to provide its suppliers real time demand information. This reduces time to market, creating value that can be shared among buyers, suppliers, and customers. In addition, since there is very little inventory in the supply channel, design and customer service problems can be resolved very quickly.

As typical for good JIT firms, Dell aggressively works with its suppliers to reduce the total inventory in the supply channel. Its suppliers typically keep less than two weeks of inventory, whereas the competitors keep nearly two to three months of inventory. This limits amount of inventory it has to sell at reduced costs when the product design changes. In order to keep inventory levels low, Dell builds long term, stable relationships with their suppliers. With certain suppliers, Dell has such a strong degree of trust it does not carry any stock inventory at all. Stock is delivered to Dell and placed right in the production line without being warehoused or inspected. Dell is able to do this because it promises its suppliers JIT-type long-term, high-volume contracts. This allows Dell to invest its capital in other facets of the organization and not have it tied up in inventory. In fact, Dell is able to turn its inventory at least two to three times that of their competitors.

The impact of reducing the variability in the system through JIT for Dell's suppliers has made production easier. Dell only ships product when it has real demand from end customers. By shrinking the size of the supply channel, the suppliers are given better information about real demand, which reduces variability, inventory, and costs. Without the demand distortion created by lengthening the supply channel, costs are reduced substantially and savings are shared across the supply-chain.

Demand Forecasts

One of the reasons Dell has been so successful is its proactive approach to forecasting demand. The better the demand information used in JIT the more effective and profitable the company (Phillips, 1998; Sanders, 1997). Dell uses a very elaborate bottom-up approach to forecasting. By doing this, sales managers can work very close with customers and access its current and future computing needs. This creates accurate demand information as to the future of the market and allows Dell to create very accurate demand forecasts. Essentially Dell treats its customers as an active member in the supply-chain. This is a component often overlooked in the development of a supply channel. To make JIT work at its best, the customers should become actively involved in the process.

Dell takes a very aggressive role to managing its supply-chain. The results of its efforts are staggering. In 12 years Dell has gone from a start-up business to a $12 billion a year business. What Dell has done is proactively apply the principles of JIT and use them in combination with advanced information technology to create its competitive advantage. How has Dell created its empire? The answer lies in its superior supply-chain design. It is a shining example of the capabilities of JIT when used properly.

Not every company can achieve these type of results, but the principles of supply-chain design can be applied to virtually any situation. Dell was in a unique situation that allowed them to build its supply-chain from scratch. Many businesses are in industries with well established supply-chains and are forced to reengineer these supply channels to gain Dell type of benefits. The remaining portion of this chapter will focus on the creation of a new JIT supply-chain and methods for reengineering existing supply-chains.

WHAT TO CONSIDER IN SUPPLY-CHAIN DESIGN FOR JIT

The difficulty in supply-chain design is deciding where to start. For many organizations the design is structured around the product design, others focus on process design, and still others around logistics design. A well designed supply-chain takes in factors from many facets of the organization to best meet the needs of the customer. A key point in the design of a JIT supply-chain is that variability must be accounted for in every organization in the supply-chain. Many organizations are able to adjust for variability within their own organization, but they fail to see the ramifications of their adjustments across the supply-chain. The impact of making changes to an individual organization in a supply-chain can often be detrimental to the supply-chain itself (see Fox, 1998).

When designing a supply-chain there is no one solution that is universally acceptable. The supply-chain for one organization may not be effective for another. In addition, the supply-chain for products within a single organization needs to be tailored to the characteristics of that particular product. However, the knowledge gained from the development of one supply-chain can be transferred to another.

Since the inception of JIT the focus has been on removing variation from the system to increase profit margins and better serve the customer. In today's marketplace extreme pressures have been placed on organizations to reduce costs, and increase customer service, making it much more difficult to remain competitive and profitable. To remain competitive, organizations are focusing on the capabilities of their supply-chain. The design of a well-built supply-chain will potentially lead to a firm gaining a competitive advantage over their competitors.

Fox (1998) and Rockhold, Lee, and Hall (1998) have point out that crucial to supply-chain design is the configuration of the manufacturing and distribution network. Associated with these decisions are the activities carried out at each site. These decisions include such things as what products are produced at the facilities, amount of inventory held, and the service responsibilities of each facility. Finally, R&D decisions must be made at a centralized location or be carried out at the each satellite facility. The issue of global supply-chain design has been a topic that has gained popularity over the past few years. Arntzen et al. (1995) discussed in great detail the modeling elements associated with Digital

Equipment Corporations' manufacturing and distribution strategies. Included in their analysis are the following elements: restrictions on local content, offset trade, taxes, duties, and duty drawbacks. All of the conditions are elements that must be considered when designing a global supply-chain. These elements have a tremendous impact on the effectiveness of JIT systems and supply-chain.

Missing from the above analysis are the elements associated with implementation and a complete cost factor analysis associated with the development of the supply-chain model. Rockhold, Lee, and Hall (1998) described a process for managing projects focusing on achieving strategic alignment of a supply-chain and the complete implementation of cost elements associated with achieving the alignment. The next section briefly describes the modeling process, cost elements associated with the model, and the related implementation issues associated the supply-chain models.

Modeling Process

To facilitate the discussion of how to align supply-chains, project teams should use analytical models to capture the appropriate elements to rationalize the decision making process. A good model includes multiple scenarios to test a wide variety of supply-chain design alternatives. Even though an organization may be implementing JIT, a wide variety of alternatives may facilitate an effective JIT supply-chain design. The results from the analytical models provide a set of alternatives for the team to evaluate.

Model Development

In the development of a model for strategic analysis proper performance metrics are needed for evaluation. To effectively manage any process, including the strategic evaluation process, the development of good metrics is critical. Both qualitative and quantitative metrics is necessary in the development of an effective model. The qualitative metrics are used to complement the quantitative performance metrics. The output from the models can be quantified into several categories. These categories include financial performance, customer service, and operational performance metrics. Key financial performance measures include, but are not limited to, return on assets (ROA), total cost, and profitability. Key customer service measures include service fill rates, customer response times, and response time to fill a customer's order. Finally operational measures include measures such as cycle time, capacity utilization, inventory level, and new product time to market.

As previously mentioned, a good model not only measures the current system to develop a baseline for comparison purposes but also evaluates multiple alternatives in the development of a new supply-chain for existing or new products. The development of new scenarios should be based on differentiating factors for both the product and process. For example, one method to differentiate supply-chains is by product and another can be based upon market characteristics.

Issues to include when considering differentiation by product include volume, distribution channels, and product design. Many products are better suited for JIT; therefore product volume levels are a very important consideration in the development of alternatives for evaluation. JIT products tend to be of larger volume and generally more standardized. The distribution channel considerations are becoming of greater importance in supply-chain design. As demonstrated by Dell, distribution channels can lead to a superior competitive advantage. Dell developed a direct model of distribution, which was very successful. Other firms, such as General Electric, have a highly integrated and well controlled system of distribution centers. One additional consideration is the choice a firm has to make in product design. Issues to consider include the complexity of the product in terms of capabilities and options and level of sophistication in design. These trade-offs add both cost and variability to the manufacturing process, which is a problem JIT advocates must deal with on a daily basis.

An additional method in which we can differentiate how firms compete is by geographic considerations. Issues to consider when competing by geographic location include whether to centralize or decentralize, and the various transportation alternatives available to the firm. There are various stages to globalization, including issues surrounding whether to centralize or decentralize distribution and production facilities (Ohmae, 1995; Schorr, 1998). Many larger JIT organizations, such as Toyota, have begun to decentralize their operations. However, in the decentralization process there are many ancillary activities that must be dealt when integrating JIT. Issues such as proximity of suppliers and frequent deliveries of shipment are just a few of the associated issues when integrating JIT. One additional aspect to consider is the transportation alternative from manufacturer to retailer and from supplier to manufacturer. The issue surrounding transportation has a tremendous impact on supply-chain performance. Firms must weigh the costs and benefits associated with using each form of transportation. Air transportation costs are reliable and speed of delivery is very good, however other forms of transportation are very cost effective.

Within each of the differentiation strategies are associated trade-offs and related costs. Relevant costs to consider in the construction of the supply-chain model and the various scenarios are related to the both the product and the geographic considerations listed earlier. Fox (1998) and Rockhold, Lee, and Hall (1998) found these costs include but are not limited to materials costs, logistics costs, regional costs, and acquisition costs (see Table 6.1).

Hewlett Packard Case Analysis

Rockhold, Lee, and Hall (1998) have applied this methodology in the personal computer industry. Hewlett Packard was able to successfully develop a global manufacturing and distribution strategy using the aforementioned modeling technique. The project team was able to determine the appropriate metric for Hewlett Packard was ROA. A subsequent model was developed to calculate the inventory levels at each site and in transit to support the various alternatives

TABLE 6.1. Relevant Cost Factors in Supply-Chain Modeling

Materials Costs
- Holding
- Stockout
- Obsolescence and perishability

Logistics Costs
- Transportation
- Tariffs
- Government Taxes

Regional Costs
- Labor
- Taxes
- Exchange rates on currency

Acquisition Costs
- Facilities
- Equipment
- Inventory
- Personnel

---

developed. After the initial model was constructed, it was validated using the existing structure that was in place.

Five scenarios were developed and subsequent sensitivity analysis was performed to rank the various alternatives. Along with quantitative metrics several qualitative metrics were used in the analysis. The qualitative factors used in the analysis included risk management and market considerations. Factors considered in risk management include amount of factories, political locality, resource availability and many others. Qualitative factors from the market side include responsiveness to orders (build to order), competitive pressures, and regional product options. Results of the sensitivity analysis and of the qualitative factors concluded that a regional build to-order strategy would be implements for the Hewlett Packard situation.

Notice that in both the Dell and Hewlett Packard cases, extremely different strategies were used that were both very successful. In the initial situation, Dell Computers used a direct model of distribution in their supply-chain. In this current case Hewlett Packard determined that a regional build to order technique was the best for its particular situation.   However, in both cases the organizations made an assessment, which was the best approach to control variability of both manufacturing and distribution. In the case of Dell since it did not have a vertically integrated structure in place it was able to develop a strong supplier alliance to meet the needs of the customer. Hewlett Packard had already established a high vertically integrated network, and the regional build-to-order strategy worked the best. In the development of both these JIT supply-chains the companies were able to successfully remove waste from the system and build an efficient system.

Not addressed in either situation is how alternatives systems developed and how they are implemented. One of the best techniques available to address the problem of simultaneously developing a product and a process has been *concurrent engineering* (CE). This technique has been successfully used in many organizations and with a large variety of products and processes. CE can now be used to develop alternatives and implementation strategies for a successful supply-chain. The remaining portion of this chapter describes CE and how to use it in the development of a successful supply-chain strategy.

WHAT IS CONCURRENT ENGINEERING?

Concurrent Engineering (CE) has become one of the techniques organizations use to achieve the simultaneous goals of reducing costs, improving quality, reducing lead time, and improving delivery reliability. One definition of CE is the systematic approach to the concurrent design of products and their related processes, including manufacture and support (Carter and Baker, 1991). The competitive priorities of speed to market and quality have forced organizations to use concurrent engineering over the traditional design methodologies.

Traditionally, product/process development occurs in a sequential or *serial engineering* design process. As presented in Figure 6.3(a), the serial engineering occurs in stages starting with conceptualization, design and prototype, and finishing in the sequence with testing and modification, manufacture, and standardization. Information in the sequence is feed forward along with the product in the design process. What distinguishes concurrent engineering from serial engineering is that all functional areas are integrated in the design process. Early on in the design process all elements of the product are considered including manufacturability, performance, reliability, cost, and quality. Unlike the serial engineering design process the various design phases in concurrent engineering are interconnected (see Figure 6.3b). Information is exchanged between various departments of the organization including purchasing, manufacturing, material handling, engineering, and any other functional area that may be impacted by the product or process design. *Cross-functional teams* (i.e., teams of multifunctional personnel) use a variety of tools to such and quality function deployment to build in customer expectations into all facets associated with the design and manufacture of products.   While the initial investment in time and effort is quite high using the concurrent engineering process the dividends received are substantial.

There are several critical stages to the concurrent engineering process. The following are some of the key elements in the concurrent engineering process (see Carter and Baker, 1991; Cooper, 1993; Nevins and Whitney 1989; Ulrich and Eppinger 1994):

FIGURE 6.3. Serial Engineering versus Concurrent Engineering

a. Serial Engineering Design Process

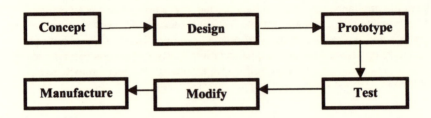

b. Concurrent Engineering Design Process

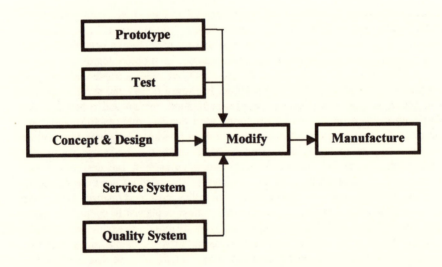

1. *Product and process analysis.* Begin by identifying problems contained within the design of both the product and process.
2. *Component analysis.* Break down the product and process into fundamental component parts and identify the interactions within and across the pieces.
3. *Alignment.* Align the requirements for the design of the product with the process and the appropriate product structure.
4. *Alternative analysis.* Develop and explore alternatives for primary product and process design.
5. *Cost analysis.* Estimate the early costs associated with product design and process selection.

6. *Time analysis*. Determine the time estimates for the adoption of the new process and design estimates.
7. *Bottleneck analysis*. Identify and alleviate any bottlenecks in the CE process.
8. *Manage and monitor*. Manage the design process with cross-functional teams and monitor their progress.

While these are not the only steps in the CE process they are some of the primary considerations. In the case of JIT, many other factors come into play that make intraorganizational CE less effective. Since JIT requires both the suppliers and distributors to be efficient, they play heavily into the design of internal products and processes. The process of CE can be extended to a third dimension and that includes the simultaneous design of the supply-chain.

CONCURRENT ENGINEERING OF THE SUPPLY-CHAIN

The traditional method of CE products and processes is insufficient to ensure a sustainable competitive advantage. To ensure companies create a competitive advantage, firms must consider the design and development of their supply-chains. Many organizations take into account the design considerations of their supply-chain; however, it is often done after the product and process have been put in place. As we discussed in Chapter 5, the supply-chain itself is only as strong as its weakest link. The failure to take into account the supply-chain from the onset of product and process may lead to disaster for an organization. Adjusting a supply-chain to fit a newly concurrent engineered product and process is better than nothing at all, but does not ensure the optimal supply-chain has been put in place.

The ramifications of not considering supply-chain design when developing the supporting product and process infrastructure can be extreme. Problems will often be encountered in product development, manufacturing launch, quality control, logistics support, and more often than not in product costs. These problems surface because of the supporting capabilities of the other members of the supply-chain do not complement a new product and or processes.

Fundamental issues arise when considering the shape and structure of the organizational supply-chain. The fundamental issues surround the analysis of both the product and process. Unlrich (1995) has made a distinction as to whether the component parts of a product are modular or integral. The distinction of modular or integral design has major ramifications as to the design of the supply-chain. In his book *Clockspeed*, Fine (1998) developed the logic of how to CE a supply-chain by using the modular and integral distinction.

An integral product architecture consists of product components performing many or multiple functions, components in close proximity, and components that are highly synchronized. In a modular design the component parts are generally interchangeable and, individually upgradeable; component interfaces are standardized; and problems are generally localized. An example of integral design is a utility vacuum cleaner which most models can be used as both a leaf blower or a straight vacuum cleaner. Computers are good examples of products that are modular in nature. Many of the parts are either interchangeable or can be easily upgraded with similar but technologically superior parts.

SUPPLY-CHAIN ARCHITECTURE

Using the concept of product architecture expanded to the supply-chain we can develop a greater appreciation of supply-chain design. The traditional trade-off in the design of the supply-chain is the make/buy or vertical integration decisions. As exhibited by Dell, a high percentage of the decisions are to contract pieces of their manufacturing to various component suppliers in the industry. However, expanding the decision horizon to include the elements of integral or modular product/process design gains a better understanding of proper supply-chain design.

According to Fine (1998) and Rosenthal (1992), proximity can be measures in terms of four dimensions. Those four dimensions are geographic, organizational, cultural and electronic:

1. *Geographic.* Roughly translated means physical distance. *Geographic proximity* is very important in the design of JIT supply-chains, because of the requirements of frequent deliveries. This dimension can be important in the design phase when integrated teams from cooperating firms are used. Physical proximity makes communication and coordination much easier.
2. *Organizational.* When suppliers and customers have vested interest and ownership rights to a particular organization, this can be referred to as close *organizational proximity.*
3. *Cultural.* *Cultural proximity* capture the elements that are common amongst nations or specific regions. This captures the elements such as language, laws and business standards, and ethics.
4. *Electronic.* Finally, *electronic proximity* refers to the exchange of information through the "information superhighway." Using tools such as the Internet, Electronic Data Interchange, and other e-commerce tools many organizations have helped to reduce speed time to market in their new products.

In contrast to an integral supply-chain, if the supply-chain is modular in nature it will feature low proximity along the four dimensions. Meaning modular supply-chains can exist over huge geographic expanses, have separate ownership control, have little commonality along culture and use very little exchange of data via e-commerce. Having little or no proximity along all of the dimensions will render a supply-chain virtually unmanageable. Dell is an example of a high modularized supply-chain. There is little or no vertical integration in Dell, and parts are virtually interchangeable.

What is the relationship between functionality of product and process, and supply-chain architecture? The relationship between product and supply-chain architecture is fairly straightforward. In general, firms, which produce modular products, tend to be aligned along the modular spectrum. Likewise firms that tend to produce integral products tend to be much closer along the dimensions of proximity. For example, interchangeable parts and a high degree of modularity characterize the personal computer industry. Of the competitors in the PC industry, those who have designed a modular supply-chain have been very successful.

Likewise those products which are highly integrated and complex tend to have a highly integral supply-chain. The best example of a integral supply-chain can be found in the automotive industry. The nature of the design and complexity of these products force the designers from many firms to be in close proximity to each other on many dimensions. As evident by the high concentration of automotive contractors and producers in the upper Midwest. This industry requires a highly integral supply-chain.

Unlike the description of the product and supply-chain architecture, the process relationship is somewhat different. Successful organizations can find themselves along any of the dimensions of the process/supply-chain matrix. There are still some observable differences. Computers usually are assembled at a very high rate and are done by one or more individuals in a work cell. In contrast, the processing of newspaper information is collected and reported from many locations dispersed across the United States and world.

The challenging part of the equation is integrating all of the elements of product, process and supply-chain architecture. While some of the activities throughout the supply-chain are still functionally separate, many of the activities overlap and are highly integrated. As Fine (1998) and others (see Carter and Baker, 1991; Cooper, 1993) have suggested, not all of the tasks must be performed using integrated development teams but there are portions that must be overlapped to ensure success. In general these authors define the overlapping elements as follows:

1. *Product development* is divided along the integral or modular architectural and detailed design choices (such as performance and capabilities).
2. *Process development* is divided along technology and equipment choices (manual vs. automation) and manufacturing system orientation (job shop vs. process flow).
3. *Supply-chain development* is divided along supply-chain architecture and infrastructure decisions. Supply-chain architectural decisions include decisions about the integral or modular nature of the product or process. Infrastructural decisions include transportation and information system decisions.

All of these dimensions are common to all supply-chains. What separates the successful supply-chains from the unsuccessful supply-chains is the degree to which they can integrate the overlapping areas. Those overlapping elements have a tremendous impact as to the firms chosen for suppliers and the types of coordination and integration elements that are used in the supply-chain. In the case of JIT organizations and supply-chains the level of success is highly dependent upon the level of integration between and within organizations. The failure to use the integrated planning teams for the overlapping coordination and planning activities will often result in the introduction of additional variability into the supply-chain. The fact that tasks and coordination of the tasks are so crucial to JIT success, the inability to account for those overlapping activities is a recipe for failure.

SUMMARY

This chapter has outlined some examples of organizations that have achieved a great deal of success because of their supply-chain design. A discussion and illustrations of how some firms have successfully used their supply-chains were also presented.

The actual design of the supply-chains varies from organization to organization and from product to product. An organization can use many tools and techniques to develop an effective supply-chain. One such method is through elaborate mathematical modeling techniques. The success of the model is dependent upon the accuracy of the inputs and the choice of the performance metrics. The models provide the integration and supply-chain design teams with additional inputs that make the decision-making process easier.

One other method that exists is through CE both product, process, and supply-chain. The methodology is much the same as in the traditional CE process, but a third dimension of supply-chain considerations are factored into the process. This helps to ensure that the supply channels and members are not ad hoc put in place to support the new or existing product lines. Whether it is in support of existing products or new products the CE process can help design an effective supply-chain.

What is clear is that the design of a supply-chain has a tremendous impact on firm profitability and success. Poorly designed supply-chains have caused some very prominent companies to struggle and become not profitable. Many organizations are not willing to take the time to factor in all of the elements of manufacturing and distribution from the onset of product design. However, research suggests that the majority of problems associated with products occur early in the design process and the later in the product life cycle of the product the more costly it is to make changes (Singh, 1996). Which would suggest that accounting for all facets of production from supplier selection through manufacturing and scheduling all the way to distribution is advantageous for competing firms. Indeed, scheduling can be an important key in making the supply-chain planning pay off. The next chapter introduces a discussion on JIT scheduling using kanbans.

REFERENCES

Arntzen, B. C., Brown, G. G., Harrison, T. P., and Traftom, L. L. "Global Supply-chain Management at Digital Equipment Corporation," *Interfaces*, 25 (1995), pp. 69-93.

Bendiner, J. "Understanding Supply-chain Optimization: From "What If" to What's Best," *APICS: The Performance Advantage*, 8 (1998), pp. 24-38.

Carter, D., and Baker, B. *Concurrent Engineering: The Product Development Environment for the 1990s.* Vol. 1, Reading, MA: Addison-Wesley Publishing, 1991.

Cooper, R. G. *Winning at New Products: Accelerating the Process from Idea to Launch*, 2nd ed. Reading, MA: Addison-Wesley Publishing, 1993.

Fine, C. *Clockspeed: Winning Industry Control in the Age of Temporary Advantage*. Reading, Ma: Perseus Books, 1998.

Fox, M. L. "Charting the Course to Successful Supply-chain Management," *APICS: The Performance Advantage*, 8 (1998), pp. 44-48.

Handfield, R., and Nichols, E. *Introduction to Supply-Chain Management*. Upper Saddle River, NJ: Prentice-Hall, 1999.

Margretta, J. "The Power of Virtual Integration: An Interview with Dell Computer's Michael Dell," *Harvard Business Review*, (March-April 1998), pp. 73-84.

Mortan, T., and Pentico, D. Heuristic Scheduling Systems: With Applications to Production Systems and Project Management. New York: John Wiley and Sons, 1993.

Nevins, J., and Whitney, D. *Concurrent Design of Products and Processes: A Strategy for the Next Generation*. New York: McGraw-Hill, 1989.

Ohmae, K. "World View: Putting Global Logic First," *Harvard Business Review*, 73 (1995), pp. 119-127.

Phillips, S. "A Total Business Systems Approach to the Supply-chain," *APICS: The Performance Advantage*, 8 (1998), pp. 54-58.

Rockhold, S., Lee, H., and Hall, R. "Strategic Alignment of a Global Supply-chain for Business Success," *POMS Series in Technology and Operations Management*, 1 (1998), pp. 123-131.

Rosenthal, S. *Effective Product Design and Development: How to Cut Lead Time to Increase Customer Satisfaction*. Homewood, IL: Business One Irwin, 1992.

Sanders, N. R. "The Status of Forecasting in Manufacturing," *Production and Inventory Management Journal*, 38 (1997), pp. 32-39.

Schorr, J.E. *Purchasing in the 21st Century*. New York, NY: John Wiley & Sons, Inc., 1998.

Singh, N. *Systems Approach to Computer Integrated Design and Manufacturing*. New York: John Wiley & Sons, 1996.

Ulrich, K. "The Role of Product Architecture in the Manufacturing Firm," *Research Policy*, 24 (1995), pp. 419-440.

Ulrich, K., and Eppinger, S. *Product Design and Development*. New York: McGraw-Hill, 1994.

7

Topics in JIT Scheduling

The intent of this chapter is to explain what role scheduling plays within an organization and in the context of JIT management. The focus of this chapter will be to present several models and research findings related to scheduling in a JIT environment.

WHAT IS THE ROLE OF SCHEDULING IN AN ORGANIZATION?

Scheduling concerns the allocation of limited resources to tasks and activities over a period of time. The competitive priorities of reducing lead time, increasing customer service, and improving delivery reliability have made organizations more of aware of their scheduling systems. There are very few organizations that have such complete control over their markets that their scheduling system is not a critical element to their organizational success. For most organizations effective scheduling is one of the ways they can gain a competitive advantage (Flynn et al., 1997). Most companies realize that effective sequencing and scheduling are necessity to survival in the marketplace (Pinedo, 1995).

SCHEDULING CONTROL FUNCTIONS

Their are many activities associated with effective scheduling systems for both service and manufacturing organizations. A good schedule not only determines the sequence of jobs to be processed on a particular machine or production line but also factors in human resources and materials. Mortan and Pentico (1993) and Pinedo (1995) view the control functions associated with scheduling are:

1. *Master production scheduling* (MPS). The MPS is created and other schedules for the timing of the necessary materials are developed in this phase of the scheduling process.
2. *Allocation of resources.* This scheduling function involves the allocating orders, equipment, and personnel to work centers to other specified locations.
3. *Sequencing and dispatching rules.* This function involves the sequencing of jobs on equipment and assembly lines. During this step of scheduling, the appropriate performance measures for the scheduling system are established.
4. *Shop floor control.* During the actual production sequence, jobs are reviewed and schedules are updated based on the status of the job. At this point in time an organization may want to consider expediting late or critical orders.

A hierarchical representation of the scheduling process is represented in Figure 7.1. The scheduling function includes both higher level strategic functions as well as shop floor control functions.

The remaining portion of this chapter focuses on current topics in JIT scheduling related to the previous scheduling control functions. Much of the literature is related to the job sequencing and dispatching rules.

MASTER PRODUCTION SCHEDULING

Most successful manufacturing organizations are able to create good MPS. Quite simply, the MPS is the central document, that determines the effectiveness of the production department and largely the organization. Developing a good MPS system is especially crucial for JIT firms where the timing of resources and material are so crucial the success of the operation.

In a "push" manufacturing environment the MPS is generated based on the effectiveness of the forecast systems. Often, these forecasts are scheduled several months in advance of when the actual production of the product is needed. This lead time gives the organization some flexibility as to when the orders for materials and inventory need to be placed since these items are often placed into inventory for large periods of time. On the other hand, a "pull" manufacturing system requires both the delivery of materials and inventories to be near perfect because of the desire to avoid storage of these items.

The major activities in MPS entail the determination of master schedule items, planning horizon, time buckets, time fences, and lot sizes. One of the big issues surrounding the MPS is how to plan for demand uncertainty. A common practice to handle this demand uncertainty is to reschedule the MPS periodically and freeze a portion of the schedule in each planning period.

Kern and Wei (1996) examined various loading methods for replanning the MPS in a JIT, capacity constrained environment. Loading allows manufacturers to control the amount and timing of MPS changes, by imposing rules on how rescheduled orders will be assigned capacity. There are three different loading

FIGURE 7.1. Hierarchical Representation of Scheduling Activities

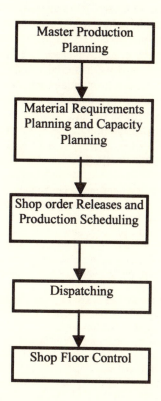

methods an organization can employ to replan the master production schedule. These three methods are level loading, front loading and back loading.

1. *Level loading.* Level loading attempts to maintain a level MPS within a specific time interval (Mortan and Pentico, 1993). To handle changes in demand uncertainty level loading makes regular, frequent updates to the MPS. Ideally, the benefits one gains from a level loading method will outweigh the effect of having to make numerous MPS changes.
2. *Front loading.* Front loading is replanning the MPS as early as possible after the frozen window subject to the capacity limitations of the system. The idea is that front loading a MPS schedule will allow the schedule to be more responsive to order changes.
3. *Back loading.* Back loading is replanning the MPS as late as possible after the frozen window subject to capacity limitations. The idea with back loading is to minimize the amount periods that are affected by the MPS changes.

The loading policy a firm adopts will have a tremendous impact on the rescheduling effectiveness of an organization. Firms must make a trade-off as to whether to continuously update the master plan (i.e., level load), adopt changes early in the process (i.e., front load), or adopt changes late in the process (i.e., back load). Under any of these situations, the loading policy is an important consideration. Since the goal for most JIT organizations is to create a level production schedule, a level loading policy would seem to be the most appropriate choice. In JIT environments, manufacturers implement many similar tactics for scheduling and operations. These tactics include rate-based scheduling, mixed model scheduling (to consume and use material evenly), frequent revisions to the schedule, and rigid time fences to maintain schedule flexibility (see Kelle and Miller, 1998). All of these tactics are in support of the level loading policy.

Kern and Wei (1996) and Kizilkaya and Gupta (1998) summarized the impact of loading policies on the replanning of the MPS. The factors they considered in their research were loading policy, size of time fence, capacity level, demand variation, and forecasting methods.

1. *Loading method.* Level loading turns out to be the most robust performance measure, reinforcing the rate based scheduling used by most organizations. This does not mean that it was necessarily the best method to use in all circumstances. If the organization is concerned with schedule instability they should consider a back-loaded policy. Alternatively, if they are concerned with potential lost sales due to demand uncertainty a front-loaded policy may be more appropriate. Finally, those who are highly concerned with the inventory control should consider the level-loaded policy to be the most appropriate.
2. *Time fences.* An optimal time fence is not of great importance to system performance. The key is to develop a "reasonable" size window. Slight variations have little or no impact on overall system performance.
3. *Capacity level.* Capacity is an important factor in system performance. To handle those uncertainties in demand and lost production having some excess capacity is an important factor for organizational success.
4. *Demand variation.* It goes without saying that JIT firms who are able to level demand out are more successful than those who cannot. As pointed out in Chapter 5, demand amplification and demand uncertainties are one of the principal reasons why JIT organizations struggle.
5. *Forecasting methods.* In any MPS for a make to stock organization, the key to developing a good schedule is based on the accuracy of the forecast. This is no different in JIT organizations as an accurate forecast helps with every dimension of system performance. In Chapter 5 this phenomenon was referred to as the bullwhip effect.

Once an organization has been able to build an accurate and effective MPS and MPS rescheduling system, then other facets of the organization need to be considered for scheduling purposes. The other facets that must be considered are the timing and allocation of the resources, and the actual machine and shop floor scheduling rules.

ALLOCATION AND TIMING OF RESOURCES

One of the big concerns regarding JIT and scheduling is the proper timing of resources to the production floor. As a result, pressure is forced upon suppliers to have shipments delivered in regular intervals and in small quantities. Ideally, the organization would like to have a shipment size that would permit it to be delivered directly to the shop floor without being stocked in inventory. The result of this increased delivery schedule has placed a tremendous amount of pressure on shipping and receiving department to perform very efficiently and effectively. Even if other organizations are able to ship the deliveries on time, if the shipping/receiving department is not able to unload the cargo on time production can be delayed.

Since most organizations' shipping and receiving departments are fairly limited on space the scheduling of the unloading of the trucks becomes an important issue for JIT firms. In fact, many organizations have adopted penalties for poor timing of shipments (Dixon, 1997; Federgruen and Moshieov, 1996). In a JIT environment the poor timing of deliveries can refer to either early or late shipments. Mukhopadhyay (1995) presented a problem that involves penalties for both early and late unloading of trucks for production. The early penalty for unloading trucks is incurred because the material must be transported and stored prior to production. The cost model for the early/late penalty in shipping/receiving is presented in Table 7.1.

As trucks arrive to the shipping and receiving area, there are often too many to just simply unload the trucks using the earliest due date rule or the first-come, first-served rule to ensure product arrives to the production floor on time. The model determines the cost associated with the early and late delivery of shipments to the production floor. In this model the cost incurred for the organization is based on the due date, unloading time and the earliness and tardiness penalties. In most instances the β is going to be greater than the earliness penalty, α.

TABLE 7.1.   JIT Delivery Scheduling Model

$$f(D) = \sum_{i=1}^{n} (w_i E_i + \beta w_i T_i)$$

where:

$f(D)$ is the total penalty cost of the JIT delivery,
n is the total number of trucks,
α is the earliness penalty,
β is the lateness penalty,
w_i is the unloading time,
E_i is the max $(0, d_i\text{-}C_i)$ is amount of earliness,
T_i is the max $(0, C_i\text{-}d_i)$ is amount of tardiness,
C_i is the completion time for truck i, and
d_i is the due date for truck i.

Mukhopadhyay (1995) developed an algorithm to determine the optimal sequence for unloading trucks (see Table 7.2). The optimal sequence algorithm is based on the decomposition method with precedence relationships developed by Szwarc (1990). An example using this sequencing algorithm is presented in Table 7.3. The results of problem in Table 7.3 using the sequencing algorithm are presented in Figure 7.2, and provides an organization with several benefits.

TABLE 7.2. Mukhopadhyay Optimal Sequencing Algorithm

Step 1   Arrange trucks in schedule in a decreasing order, such that

$$w_1 \geq w_2 \geq \ldots w_n,$$

if $w_i = w_j$ then truck with the lower due date is placed first.

Step 2   For each pair of sequential trucks calculate q_{ij} where $i < j$

$$q_{ij} = d_i - w_i - w_j + \frac{d_i - d_j}{w_i - w_j} w_j, \text{ where } w_i \neq w_j.$$

Step 3   Set initial sequence by following these rules
   - Rule 1: If $d_i \leq d_j$, then i precedes j.
   - Rule 2: If $d_i - d_j \geq w_i - w_j$, then j precedes i.
   - Rule 3: If $0 < d_i - d_j < w_i - w_j$, then j precedes i for the $t < q_{ij}$ and i precedes j for $t > q_{ij}$.
   - Where t is the starting time of unloading of the pair (i,j). An triangular matrix for $0 < i < j < n$ is developed from these rules by placing a (+) for rule 1, (−) for rule 2, and q_{ij} value for rule 3.

Step 4   Decompose into smaller blocks
   - Rule 1: If cell (i,j) is a q_{ij} then jobs i and j go into one block
   - Rule 2: All jobs k such that cells (i,k) or (k,j) are q_{ij} go into the same block
   - Rule 3: Any job k where entries in column k and row k are a "+" or "−" is a single job block
   - Rule 4: If all the entries in the matrix are "+" in column i and "−" in row i, then job i is the first job of the block.

Step 5   Optimal sequencing of the blocks
   - The optimal arrangement is found by sequencing the blocks found in Step 4 nondecreasing order.

TABLE 7.3. Problem Statement and Answer for Optimal Sequencing Algorithm Example

---

Problem Statement: Suppose we have a situation where there are eight trucks scheduled to be unloaded in the morning. The trucks are scheduled to start unloading at 7 AM for production that starts in the facility at 8 AM. The estimated earliness and tardiness penalty are $\alpha = 1$ and $\beta = 3$. Since unloading starts 60 minutes prior to production the due date times start at t = 60 minutes.

	A	B	C	D	E	F	G	H
w_i	30	25	15	15	15	15	10	10
d_i	65	60	65	70	120	130	60	70

Answer: If we follow the typical unloading procedure of unloading by the earliest due date sequence the schedule would be as follows: {G, B, C, A, H, D, E, F}. We will see in the solution that the optimal sequence is actually quite different than the commonly used earliest due date rule. The solution using the optimal sequencing algorithm is:

Step 1   Arrange trucks in schedule in a decreasing order of their unloading times. {A, B, C, D, E, F, G, H}

Step 2   Generate the upper triangular matrix for the eight trucks in sequence. The matrix is presented in Figure 7.2.

Step 3   Decompose the jobs into blocks by applying rules under Step 4. This example can be decomposed into the following blocks: {(A, G), B, C, D, E, F, G, H}

Step 4   Apply Step 5 of the heuristic to obtain optimal sequence of the unloading schedule. If all the blocks have single jobs the sequence is complete. This example yields the following sequence: {B, (A, G), C, D, H, E, F}

Step 5   Sequence jobs within the remaining blocks. Sequence within the blocks by comparing all possible combinations and determine the feasible set. For this case A cannot precede G at time t = 25 which is the starting point of the second job. Therefore the optimal sequence is: {B, G, A, C, D, H, E, F}

First, it provides an optimal schedule for unloading of trucks. Second, the schedule can be generated and regenerated in a matter of seconds. This could be beneficial if a truck is delayed in route. The planner can easily regenerate the schedule for the shipping and receiving department. Third, and finally, as with

FIGURE 7.2. Triangular Matrix for Optimal Sequence Example

A	B	C	D	E	F	G	H		
	+	–	–	–	–	28	–		A
		–	–	–	–	–	–		B
			–	–	–	+	–		C
				–	–	+	–		D
					–	+	+		E
						+	+		F
							+		G
									H

any computer generated model the planner can generate a variety of "what if" type questions based on current conditions in the environment.

There are many other resources an organization must consider when developing their schedules for production. Other elements an organization must consider are the movement of materials through the manufacturing process and human resource scheduling. Usually material movements in a JIT environment is controlled by kanbans (see Chapter 4), but many organizations use automatic storage and retrieval systems (AS/RS) to support their material movement efforts. Lee and Kim (1995) presented several two-stage models for AR/RS scheduling in a JIT environments.

After the MPS and associated resources and materials have been scheduled the question of how to actual schedule jobs onto the various production lines arise. This question was easy to answer in the days of Henry Ford's production line where the entire product line was identical and the organization had no competition. Customers are demanding greater product variety and much quicker than ever before placing a huge emphasis on the effectiveness of the scheduling system (see Roman, 1999). In particular, the issue of how we sequence jobs through the production system has become of great importance (see Aytug and Dogan, 1998).

SEQUENCING AND DISPATCHING RULES

Sequencing and *dispatching* deals with the actual processing of jobs through the manufacturing system. Once job sequences have been developed the actual execution of the schedule depends on the many factors within the production environment. Several authors have addressed the critical components associated with work station scheduling. A few of the principles are listed below (Chase, Aquilano, and Jacobs, 1998; Gaither, 1992; Grout, 1998) are listed below:

1. There is a direct equivalence between effective scheduling systems and the bottom line.
2. The speed and consistent flow of materials and product through the shop measure the effectiveness of any schedule.
3. Schedule jobs as a string, with process steps back to back.
4. Once started, a job should not be interrupted. Interruptions add unnecessary variability to any process.
5. Speed of flow is most efficiently achieved by focusing on bottleneck work centers and jobs.
6. Reschedule frequently. As shown earlier, rescheduling leads to a more effective production system. If our environment is dynamic so should our schedules be dynamic.
7. Obtain feedback every day on jobs that are not completed by work centers.
8. Match work center input information to work center capacity target levels. Never expect a schedule to be effective if it places unrealistic assumptions about capacity of work centers.

In essence, what all of these characteristics of work station scheduling point toward is that there is variability present in all scheduling systems. Managers must realize that there is variability and make appropriate plans to cope with its effect on the system. The effective organization is the one that aggressively attempts to remove that variability from their system and is able to adjust their schedule accordingly, this is the purpose of JIT and JIT scheduling systems.

Scheduling Rules

One of the biggest issues in multiproduct, multiline JIT scheduling is how should a production worker choose the next job for processing based on available information. The essential question is which job to process out of a preceding queue. In practice the first-come, first-served rule is commonly used (Hum and Lee, 1997; Lummus, 1995). The use of this scheduling rule is quite common because its use is very easy and requires little understanding of the actual system. The question arises, then, what is the most effective scheduling rule to use in a JIT manufacturing environment?

To answer this question, we can investigate the impact of changing the scheduling rules on a typical JIT system. Lummus (1995) investigated the impact of various sequencing rules on the performance of a JIT system with multiple lines and multiple work stations given variable setup times and processing times. The performance measures used in the experiment were total process time and process utilization.   The purpose of the research was to examine the effects of sequencing on a final assembly process with multiple products and variable setup and processing times. The three sequencing alternatives examined were:

1. *Toyota production rule.* This rule tries to build consistency in the scheduling process by producing the same product on the same day of the each month. A typical

sequence using the Toyota Production rule might look like the following: (AAAAABAAAAAB).

2. *Least changeover rule* (LCR). This scheduling rule is typically employed by firms using a JIT system. The LCR minimizes the amount of setups in the process. This can often be found in industries where setups are sequence dependent and poor scheduling leads to a large increase in set-up time. A typical sequence using these production rules might look like the following: (AAAAAAAAAABB).

3. *Demand driven rule.* The finished goods that are removed from the production system determine the demand driven sequence. The production sequence is entirely driven by the rate and flow of customer demand.

All of these rules are commonly used by JIT organizations in their work station/assembly line scheduling. The results of the study indicates that large increases in set-ups and processing times at any production center can cause large disruptions to a JIT production system. Kanbans are one of the mechanisms developed to help minimize the impact of these disruptions, however the results indicate that when this ripple occurs the system performance will decline. The end result is that customer orders may be delivered late. In addition, set-up times have a direct impact on the performance of a JIT system. Supporting the notions that setup time in one of the prime considerations of any JIT system (Shingo, 1981). Finally, in most instances the demand driven sequencing rule performed better than the other sequencing rules. Indicating, JIT firms let demand drive their schedules more than the adoption of some scheduling policy.

In a similar study, Hum and Lee (1997) investigated the impact of scheduling rules under various JIT production scenarios. Unlike the Lummus (1995) study, this research attempts to examine the impact of the scheduling rules within the system. It should be noted that they tested a serial line as opposed to a multiline JIT production system. In this serial line the final assembly station operated on a predetermined schedule based on demand. The performance measures used for this study were total time, process utilization, average number of stoppages, and duration of the stoppage. Hum and Lee (1997) tested the following rules in their simulation study:

1. *First-come, first-served* (FCFS). The FCFS simply processes job as they arrive in the preceding queue. This is the most common scheduling rule used by JIT organizations.

2. *Shortest processing time* (SPT). The SPT chooses those jobs in the queue with the lowest work completion time.

3. *Number of kanbans* (NK). The NK priority to producing the job that has the greatest number of kanbans waiting at the work center.

4. *Ratio of kanbans* (RK). The RK rule is based on the number kanbans waiting for processing relative to the total number of kanbans in the system for that particular part type. The RK rule gives priority to the jobs that have the highest ratio in the system.

The results of the Hum and Lee (1997) simulation have several implications for the use of JIT scheduling rules within a JIT process. The first the FCFS rule performs the poorest under all system conditions. Not only does it perform poorly under different system distributions, but it is also very sensitive to changes in both the numbers of kanbans and increases in product mix. The NK, RK, and SPT rules operate about equally effectively under any system operation conditions. As the setup times increase in the systems, the SPT rule operates far and above any of the other processing rules. The introduction of the number of kanbans reduced the total amount of stoppages and total time in system, but only to an upper limit. Once a system becomes saturated, adding additional kanbans has no additional impact on the system. However, if an organization is using the FCFS rule the addition of kanbans will have dramatic impact on time and the performance measures. Finally, the addition of products greatly impacted the amount of stoppages and time in system. The FCFS rule operates extremely poor when additional products are added in the JIT serial line. The summary of results for the Hum and Lee (1997) study are listed in Table 7.4.

While the Lummus study (1995) indicates the most robust scheduling system for the final assembly area is to operate under a demand driven schedule, the Hum and Lee (1997) study indicates the most effective scheduling system within the JIT processing centers is very much rule dependent. The results of both studies indicate that the scheduling rules an organization uses has an impact on the effectiveness of the organization. Some common themes begin to arise from the literature involving scheduling literature. First, the common used FCFS rule used by most JIT organizations appears to be quite ineffective. Second, the tighter the production process the worst the FCFS rule works. Finally, various production scenarios and parameters such as processing time and setup time have an impact on which scheduling system to use.

All of these functions from MPS scheduling to shop floor control and dispatching rules, have a tremendous impact on manufacturing system performance. It is quite clear that a firm should not arbitrarily adopt a scheduling system. It is also clear that the use of the scheduling system is important to match production output to demand in manufacturing and, as it turns out, is even more crucial in service systems (see Duclos, Siha, and Lummos, 1995).

TABLE 7.4. Results of Scheduling Rules on System Performance

| System Conditions | Scheduling Rule Used | | | |
	FCFS	SPT	NK	RK
Static Processing Time	Poor	Best	Good	Good
Exponential Processing Time	Poor	Good	Best	Good
Normal Processing Time	Poor	Good	Good	Best
Setup Times	Poor	Best	Good	Good
Product Mix	Sensitive	Good	Good	Good
Number of Kanbans	Sensitive	Good	Good	Good

JIT SCHEDULING IN JIT SERVICES

Scheduling in a service environment takes on a whole new dimension over and above that of a manufacturing environment. In JIT manufacturing the coordination of materials, resources, and personnel are formulated based upon the accuracy of either the forecasts or orders on hand. The success of JIT in manufacturing largely relies on the ability of the organization to adjust their schedule to meet demand and changes in demand.

In contrast, in a service organization the demand is often very unstable and unpredictable. For JIT organizations this is a very hard obstacle to overcome because of the desire to create uniform scheduling for the system. Indeed, capacity is often determined by the amount of people in the system.

Scheduling Personnel in Services

The biggest challenge for most service organizations is the scheduling of personnel to meet changes in variable demand patterns. Several techniques have been used to level demand for service organizations such as appointment systems, to vary workforce strategy to adjust capacity, and the emergency workforce strategy (Gaither, 1992; Lee and Kim, 1994; Schniederjans, 1993, pp. 231-257).

1. *Appointment systems.* In an attempt to control the flow of demand in the service organization appointments are used to schedule demand. This is a very common practice in professional services such as the medical profession. The goal is to control the flow of demand to match available capacity.
2. *Vary workforce strategy.* This strategy controls the available capacity of the service by regulating the amount of employees in the system. Again as in the manufacturing system the effectiveness of this method is largely based on the effectiveness of the service forecasts system.
3. *Emergency workforce strategy.* This system involves using emergency workforce to handle surges in demand. This is often referred to as on-call staff.

Adopting one of these simple scheduling methods does not ensure a JIT scheduling system will be effective. There are too many constraints in the environment that makes these policies ineffective for all organizations.

To examine how scheduling affects a service delivery system, it is useful to examine the types of scheduling problems that exist. Essentially personnel scheduling can be classified into three different categories: days-off, shift, or labor tour scheduling problems (Baker, 1976). *Days-off scheduling* is concerned with the designation of employee non-work days across a planning horizon. *Shift scheduling* determines the start and stop times and often the determination of breaks. *Labor tour scheduling* involves assigning employees to daily shifts that efficiently satisfy demand for labor, yet allow sufficient time for rest between shifts (Bechtold and Busco, 1995; Jacobs and Brusco, 1996). The majority of the service industries are faced with the labor tour scheduling problem.

The full-time staff tour problem is ideal for JIT service organizations who are able to level the flow of demand through their service organization. The mixed part-time and full-time tour scheduling problem is better suited for JIT service industries that use part time workforce to handle large surges in demand variability. In either case the tour scheduling provides the organization the flexibility to handle varying patterns of demand.

The goal with any labor scheduling schedule is to minimize cost and still have adequate coverage to meet demand. Since the initial model was developed mid-1950s, many computer algorithms and solutions have been developed (Bechtold and Showalter, 1987; Brusco and Jacobs, 1993, Glasserman and Wang, 1998). Bechtold and Brusco (1995) developed algorithms to efficiently handle the full-time scheduling problems (see Table 7.5). The heuristic is a simple two-stage method in which: (1) a rule-based procedure that promotes the utilization of the earliest and latest shift start times is used to solve the shift scheduling problem on a daily basis, and (2) an optimal days off algorithm is used to solve the days off problem for each starting time (Bechtold, 1988).

TABLE 7.5. Bechtold and Brusco Full-Time Tour Scheduling Heuristic

---

Step 1 $Y_{tj} = 0$ for $j = (1$ to $7)$

Step 2   Select first working day of each week (m) that meets the following re-
          quirements (all ties broken arbitrarily):
          - If $X_{t,m-3} > \max(X_{t,m-2}, X_{t,m-1})$, move to step 3, or
          - If $X_{t,m-3} > X_{t,m-2}$, move to step 3, or
          - Choose and $m = j$, $1 \leq j \leq 7$ and move to step 3.

Step 3 $Y_{tm} = Y_{tm} + 1$ and $X_{tj} = \max(X_{tj} - 1, 0)$ for $j = (m$ to $m + 4)$. If $X_{tj} = 0$ for
          $j = 1$ to 7, stop. Otherwise return to step 1.

Step 4   Once the algorithm stops for $t = (1$ to $h)$, the total number of employ-
          ees scheduled is:

$$E = \sum_{t=1}^{h} \sum_{j=1}^{7} Y_{tj}$$

where:

E is the total number of employees scheduled,
Y_{tj} is equal to the size of the workforce assigned to work week,
j is the starting day for the work week,
t is the starting hour for work shift,
s is the number of possible shifts during the day,
m is the starting day of the work week, and
$X_{t,j}$ is the number of employees assigned to the shift beginning in hour t on day j.

The solution to this heuristic procedure determines the optimal tour schedule that minimizes the total cost of the scheduling problem. While the solution of such heuristics in the past was very slow and cumbersome, the invention of software programs such as spreadsheets and linear programming software has permitted these problems to be solved quickly. For most service organizations employing JIT, these heuristics enable the organization to remove any waste in the system caused by poor scheduling (see Sharadapriyadarshini and Rajendran, 1997; and Wang, Wang, and Ip, 1999).

The scheduling of personnel in a JIT service firm represents the most challenging system to schedule and control (see Balkrishnan, Kanet, and Sridharan, 199; Dove, 1999; Dixon, 1997; Nakamura and Schroeder, 1998; Ogan and Heitger, 1999; and Porter, 1997). Certainly the scheduling of personnel is not the only concern for service organizations, but is the most difficult and important (see Kevin, 1998; Larson, 1998; and Power and Sohal, 1997). Other facets such as materials and inventories need to be scheduled in services as well as in manufacturing environments.

SUMMARY

The purpose of this chapter was to outline the various issues related to JIT and scheduling. As was presented, scheduling entails everything from setting the MPS correctly to actual shop floor control mechanisms. This chapter provided readers with related scheduling issues and specific methods to schedule various elements of the process.

Scheduling in today's manufacturing environment is a tremendously difficult task to undertake. The performance of the system depends on the effectiveness of scheduling of parts and product through the system and the scheduling of associated resources and materials. As a central, guiding document to the organization, the MPS has a very important role in the effectiveness of a production system. While it is easy to set an original MPS, it is very difficult to update the MPS as the demand schedule changes. The policy an organization uses to reschedule their MPS will largely determine the effectiveness of the entire scheduling system. In a JIT environment this issue becomes even more critical as there is no excess inventory to protect against poor scheduling policies. In the ideal JIT system the resources and materials move directly from the shipping/receiving area to the production floor as work in process. Accordingly, the methods an organization use to schedule the timing of these resources have a tremendous impact on the system performance. Another crucial element is the actual scheduling of the products and materials through the production floor. The final focus of a JIT scheduling system is to ensure the proper timing of finished goods with consumer demand. As competition continues to grow, the scheduling systems of JIT organizations will become more complex and integrated with other organizations. This will force organizations to be good at all facets of scheduling to be competitive.

Clearly the elements of JIT scheduling also apply to human resources. The difficulty arises when trying to schedule personnel to meet customer demand.

However, there are certainly more issues to deal with than just scheduling human resources in JIT firms. In Chapter 8 the book concludes with a discussion of the current issues concerning this most critical area for JIT success.

REFERENCES

Aytug, Haldun, and Dogan, A. "A Framework and a Simulation Generator for Kanban-controlled Manufacturing Systems," *Computers & Industrial Engineering*, 34 (1998), pp. 337-361.

Baker, K.R. "Workforce Allocation in Cyclical Scheduling Problems: A Survey," *Operational Research Quarterly*, 27 (1976), pp. 155-167.

Bechtold, S. E. "Implicit Optimal and Heuristic Labor Staffing in a Multi-Objective, Multi-locational Environment," *Decision Sciences*, 19 (1988), pp. 353-373.

Bechtold, S. E., and Brusco, M. J. "Microcomputer-Based Working Set Generation Methods for Personnel Scheduling," *International Journal of Operations and Production Management*, 15 (1995), pp. 63-74.

Bechtold, S. E. and Showalter, M. J. "A Method for Labor Scheduling in a Service Delivery System," *Decision Sciences*, 18 (1987), pp. 89-107.

Balakrishnan, Nagraj, Kanet, John J. and Sridharan, Sri V. "Early/tardy Scheduling with Sequence Dependent Setups on Uniform Parallel Machines," *Computers & Operations Research*, 26 (1999), pp. 127-137.

Brusco, M. J., and Jacobs, L. W. "A Simulated Annealing Approach to the Cyclic Staff-Scheduling Problems," *Naval Research Logistics*, 40 (1993), pp. 69-84.

Chakravorty, S. S. and Atwater, J. B. "A Comparative Study of Line Design Approaches for Serial Production Systems," *International Journal of Operations & Production Management*, 16 (1996), pp. 91-108.

Chase, R. B., Aquilano, N. J., and Jacobs, F. R. *Production and Operations Management*, 8 th ed., Boston: Irwin-McGraw-Hill, 1998.

Dixon, Lance "Got a Problem? Get JIT II," *Purchasing*, 123 (1997), pp. 31-32.

Dixon, Lance "JIT II: The Ultimate Customer-Supplier Partnership," *Hospital Management Quarterly*, 20 (1999), pp. 14-21.

Dove, Rick "Creating & Communicating Agility Insights," *Automotive Manufacturing & Production*, 109 (1997), pp. 18-20.

Duclos, L., Siha, S. M., and Lummus, R. R. "JIT in Services: A Review of Current Practices and Future Directions for Research," *International Journal of Service Industry Management*, 6 (1995), pp. 36-52.

Federgruen, A., and Mosheiov, G. "Heuristics for Multimachine Scheduling Problems with Earliness and Tardiness Costs," *Management Science*, 42 (1996), pp. 1544-1564

Flynn, B. B., Schroeder, R. G., Flynn, E. J., Sakakibara, S., and Bates, K. A. "World-Class Manufacturing: Overview and Selected Results," *International Journal of Operations and Production Management*, 17 (1997), pp. 671-685.

Gaither, N. *Production and Operations Management.* 5th ed., Fort Worth: The Dryden Press, 1992.

Glasserman, Paul and Wang, Yashan "Leadtime-Inventory Trade-offs in Assemble-to-order Systems," *Operations Research,* 46 (1998), pp. 858-867.

Grout, John R. "Influencing a Supplier Using Delivery Windows: On the Variance of Flow Time and On-time Delivery," *Decision Sciences,* 29 (1998), pp. 747-456.

Handfield, R., and Nichols, E. *Introduction to Supply-Chain Management.* Upper Saddle River, NJ: Prentice-Hall, 1999.

Hum, S. H. and Lee, C. K. "JIT Scheduling Rules: A Simulation Evaluation," *Omega: International Journal of Management Science,* 26 (1998), pp. 381-395.

Jacobs, L.W., and Brusco, M. J. "Overlapping Start-Time Bands in Implicit Tour Scheduling," *Management Science,* 42 (1996), pp. 1247-1259.

Kelle, Peter, and Miller, Pam Anders "Transition to Just-In-Time Purchasing: Handling Ounce Deliveries with Vendor-purchaser Co-operation," *International Journal of Operations & Production Management,* 18 (1998), pp. 53-66.

Kern, G. M., and Wei, J. C. "Master Production Rescheduling Policy in Capacity-Constrained Just-In-Time Make-To-Stock Environments," *Decision Sciences,* 27 (1996), pp. 365-387.

Kevin, Z. "Tailored Just-in-Time and MRP Systems in Carpet Manufacturing," *Production & Inventory Management Journal,* 39 (1998), pp. 46-51.

Kizilkaya, Elif, and Gupta, Surendra M. "Material Flow Control and Scheduling in a Disassemble Environment," *Computers & Industrial Engineering,* 35 (1998), pp. 93-97.

Larson, Paul "Carrier Reduction: Impact of Logistics Performance an Interaction with EDI," *Transportation Journal,* 38 (1998), pp. 40-48.

Lee, M. K., and Kim, S. Y. "Scheduling of Storage/Retrieval Orders Under a Just-In-Time Environment," *International Journal of Production Research,* 33 (1995), pp. 3331-3348.

Lummus, R. R. "A Simulation of Sequencing Alternatives for JIT Lines Using Kanbans," *Journal of Operations Management,* 13 (1995), pp. 183-191.

Mortan, T., and Pentico, D. *Heuristic Scheduling Systems: With Applications to Production Systems and Project Management.* New York: John Wiley and Sons, 1993.

Mukhopadhyah, S. K. "Optimal Scheduling if Just-In-Time Purchase Deliveries," *International Journal of Operations and Production Management,* 15 (1995), pp. 59-69.

Nakamura, Masao Sakakibara, and Schroeder, Roger "Adoption of Just-In-Time Manufacturing Methods at U. S. and Japanese-owned Plants: Some Empirical Evidence," *IEEE Transactions on Engineering Management,* 45 (1998), pp. 230-241.

Ogan, P., and Heitger, "Alphabet Soup: Good for You or an Indigestible Stew?" *Business Horizons,* 42 (1999), pp. 61-69.

Porter, Anne Millen "The Problem with JIT," *Purchasing,* 123 (1997), pp. 18-20.

Pinedo, M. *Scheduling: Theory, Algorithms, and Systems.* Englewood Cliffs, NJ: Prentice Hall, 1995.

Power, Damien J., and Sohal, Amrik S. "An Examination of the Literature Relating to Issues Affecting the Human Variable in Just-In-Time Environment," *Technovation*, 17 (1997), pp. 649-667.

Roman, Leigh Ann "Just-In-Time: Methodist Saves Big with More Deliveries," *Memphis Business Journal*, 20 (1999), pp. 3-6.

Schniederjans, M. J. *Topics in Just-In-Time Management.* Boston: Allyn and Bacon, 1993.

Sharadapriyadarshini, B., and Rajendran, Chandrasekharan "Heuristics for Scheduling in a Kanban System with Dual Blocking Mechanisms," *European Journal of Operational Research*, 103 (1997), pp. 439-453.

Shingo, S. *Study of "Toyota" Production System From Industrial Engineering Viewpoint.* Tokyo: Japan Management Association, 1981.

Wang, Wei, Wang, Dingwei, and Ip, W. H. "JIT Production Planning Approach with Fuzzy Due Date OKP Manufacturing Systems," *International Journal of Production Economics*, 58 (1999), pp. 20-31.

8

Topics in JIT Human Resource Management

This chapter provides an overview of a number of advanced human resource management topics as they relate to the practice of JIT. The purpose of the chapter is to provide a current understanding of how pressing human resource management issues are being dealt with in the context of a JIT operation.

INTRODUCTION

Human resource management (HRM) can be defined as the collection of activities of attracting, developing, and retaining people with the necessary knowledge and skills to achieve an organization's objectives (see Luthans 1998, pp. 16-17). Because so much depends in a JIT operation on the personnel that run the business, HRM is a critically important area for JIT success. Recent research on the importance of HRM activities reveals a close connection with the principles of JIT management. In a survies by Deshpande and Golhar (1995) and Deshpande, and Golhar (1996) HRM managers in JIT manufacturing operations were asked to rank the importance of various HRM activities. Those rankings are presented in Table 8.1. The top five HRM activities are themselves essential ingredients to basic JIT principles discussed in Chapter 1.

Research has shown that JIT can improve organizational operations in many of the functional activities in HRM. Current research has shown that the implementation of JIT can improve management and worker relations, improve empl-

TABLE 8.1. Top 10 Ranked HRM Activities of JIT Manufacturing Firms

HRM Activities	Overall Ranking by HRM Managers*
Communication	1
Employee participation initiatives	2
Training for new employees	3
Training to improve team effort	4.5
Continuous training programs	4.5
Job security for employees	6
Wages	7
Collective responsibility	8
Training to improve quantitative skills	9
Job enlargement	10

*Original rankings from study recomputed based on combined union and non-union HRM manager rankings.

oyee job skills and problem solving abilities, reduce the number of grievances filed by workers against management, can reduce labor turnover, increase worker morale, and improve communications internally (i.e., in the organization) and externally (i.e., between the organization and its customers and suppliers) (see Yasin and Wafa 1996; Yasin, Small, and Wafa, 1997).

To ensure operations are successful, HRM managers seek to obtain the type of personnel that have the characteristics needed for JIT operations. In the survey by Deshpande and Golhar (1995) some of these workforce characteristics and their relative rankings where identified. The characteristics and rankings are presented in Table 8.2. The HRM activities in Table 8.1 and the characteristics listed in Table 8.2 are commonly seen in almost every HRM study on JIT operations. Research has also revealed a number of HRM issues that cause JIT implementation problems (see Adair-Heely, C., 1991; Yasin, Small, and Wafa 1997; Deshpande, Golhar, and Stamm, 1994). Used in researching the difficulties in implementing JIT, Table 8.3 contains a listing of HRM issues that were shown to be statistically significant JIT-related problem factors. It is interesting to contrast the JIT implementation problems listed in Table 8.3 with those activities and characteristics on Tables 8.1 and 8.2. It is no coincidence that HRM managers are performing activities and looking for characteristics in personnel that will solve most of the problems in implementing JIT operations.

TABLE 8.2. Important JIT Workforce Characteristics and Rankings

Workforce Characteristics	Overall Ranking by HRM Managers*
Concern for organization's success	1
Ability to inspect work	2
Ability to work in teams	3
Worker flexibility	4
Self-disciplined	5
Communication skills	6
Multi-skilled workforce	7

*Original rankings from study recomputed based on combined union and non-union HRM manager rankings.

TABLE 8.3. JIT Implementation Problem Factors and Rankings

Implementation Problem Factor	Overall Ranking*
Worker resistance	1
Top management lack of support	2
Lack of communication between management and workers	3
Top management resistance	4
Lack of formal training for managers	5
Lack of formal training for workers	6
Unionized workers	7
Problems with machines and plant layout	8

*Overall rankings based on statistical loading factor significance.

The purpose of this chapter is to discuss several of the more current re-search studies focused on HRM issues in JIT operations. Research findings suggesting ideas and strategies to overcome implementation problems are pre-sented to help JIT managers know how HRM departments can aid in maximiz-ing the benefits of JIT. The specific HRM subjects addressed in this chapter in the following sections include: teamship, learning, empowerment, labor relations, and layout design.

JIT AND TEAMSHIP

A well recognized characteristic of JIT operations is the use of teams of people on the shop floor (see Heeley, 1991, pp. 35-55). Teams encourage employee participation, help to empower employees to overcome work restrictions, help to motivate new ideas that improve production, and improve the quality of work life.

TABLE 8.4. Stages of JIT Teams Development

Stage	Description of Development Efforts
1. Formation	Choose team members, team leader, and state goals and objectives for the team. Define roles and responsibilities of members, including any ground rules on attendance, decision-making methods, and scope of what is to be decided.
2. Storm	Encourage team members to brainstorm, and take risks in what they say to each other. Help to bring out shy members by challenging them with questions and asking for their opinions. This period of awkwardness will help to mold the members into a team and get comfortable and honest with each other. Hopefully a number of very good ideas may start originating from the team at this stage. Regardless of how odd the ideas may sound, record and keep tract of them.
3. Normalize	Take the ideas that come from Stage 2 and start trying to build a consensus for the best ones. Maybe use ranking methods to choose candidates or narrow down the list. Remember that you are not only improving a process or solving a problem, but you are building a team that can come together to deal with what ever issue they are charged to do. During this stage team members will learn the unique differences in the contributions each member can offer.
4. Performance	At this stage, the team is fully connected with each other. They have learned the strengths and weaknesses of each other and utilize each appropriately to achieve team goals.

Some teams of employees are used to improve production processes, while other teams are used to handle specific tasks (i.e., solve a quality control problem). Heeley (1991, pp. 44-45) suggests that there are four stages in team development similar to those presented in Table 8.4.   To make the team approach successful, hinges on JIT education. Teams must be taught the basics of JIT and must be trained to participate as a team. This education sometimes entails basic JIT instruction by educators from local colleges or consultants. The education usually involves the use of JIT decision-making tools (e.g., fishbone charts, statistical process control, etc.), and group decision-making approaches, and building communication skills.

Absenteeism

Regardless of how well the development of JIT teams is undertaken, a variety of problems can prevent its successful implementation. One very serious factor that prevents the success of teams is absenteeism. If the team members are absent, there is no team.

One advance in recent research (see Conti, 1996) suggests that one way to help minimize the negative impact of absenteeism in JIT teams is with variable manning. *Variable manning* allows changes in the actual production schedule (i.e., the MPS) to meet the variable workforce size on a daily or weekly basis. The way in which variable manning is implemented is to set up production cells that are manned by a single team. The team is allowed to reschedule work in the MPS to better match up with the variable labor team pool that changes each day. The JIT teams are workers with a single job classification and cross-trained to perform all jobs, including final tests. This permits each JIT team to produce the entire product. This approach also eliminates most of the need for standby workers, who might not have all the cross training necessary to handle all work task assignments in a fast-paced JIT production cell.

Once the variable manning JIT teams are trained, care is taken to examine prior absenteeism in planning labor needs and production requirements. Historic patterns of attendance, vacations, and maintenance requirements are all including in the planning effort to develop a HRM plan that fits the production requirements of the MPS.

As the JIT teams start operating, daily mixed-model scheduling permits the teams to adjust the workload if the necessary to match number of workers are present to complete an expected quota. Workers who show up unexpected can be transferred to production cells with unexpected absenteeism.

Conti (1996) reported that the application of this JIT team approach of variable manning was successfully applied to a commercial appliance manufacturer. The firm eliminated their standby employees, improved worker morale, improved product quality, increased MPS scheduling reliability for on-time shipments, and improved productivity. Since the JIT teams were made up of full-time employees, it was also observed that they generated more team cohesion and a reduction in overall absenteeism.

Legal Limitations

Another interesting issue that can negatively impact JIT teams are the legal limitations to self-directed work teams in production planning and control (see Abraham and Spencer 1998). The 1935 *National Labor Relations Act* (NLRA), Section 8(a)(2) can be interpreted as prohibiting the establishment of any self-directed work group or team where employee participation is required. The *National Labor Relations Board* and the U.S. Supreme Court have in several court cases agreed with this interpretation. What the research (see Abraham and Spencer, 1998; Gould, 1996; Price, 1995) suggests is that JIT teams (or any quality circle) can be in violation of law. The teams violate the law if their intent is to dissuade the formation of a union or to encourage employees to get out of a union. Since most JIT teams have as their intent the desire to improve productivity or make use of employee ideas, the best legal judgment is that on "intent," JIT teams do not violate the law. Making sure that all members, and even unions, are clearly informed as to the JIT team's intent is good advise against future suit. Unfortunately, intent is not the only possibility of violation of the law. The NLRA can also be violated if management interferes or dominates the JIT team by: (1) conducting the team meetings, (2) calling or scheduling the team meetings, (3) setting the agenda for the team meetings, or (4) choosing the team members to attend the meeting. How can these common team activities not be performed by management? Abraham and Spencer (1998) have suggested that a supervisor or manager not be a part of the JIT team. This avoids the perception of control of the team by management. If a supervisor or manager must be a part of the JIT team, then he or she should not be the team leader. The teams should be empowered to make their own decisions and recommendations without the direct or even perceived indirect influence of management. While the JIT team may keep management informed as to their efforts for the organization, management should not be given any veto power over the JIT teams actions.

This legal problem concerning JIT teams is not yet a serious issue, but can be viewed as potentially critical in the future. It can be viewed as an advanced topic that managers should be aware of now. Managers should prepare themselves now to deal with the inevitable future legal labor claims.

JIT AND LEARNING

One of the areas of responsibility in HRM is training. From training it is hoped that the eventual outcome is "learning." One of the fundamental principles necessary for JIT to be successful is that management finds problems that JIT helps to reveal, and corrects them to improve operations efficiency. These problems that are identified and resolved are a part of the "waste" that JIT must have in order to make it a successful approach to management. In other words, a JIT operation must be able to "learn from mistakes" in order to improve operations.

Recent theoretical advances (see Ocana and Zemel, 1996) on learning in JIT systems has shown that learning conditions occur at a time that may be more costly than needs to be.   That JIT, in revealing a problem like an inventory shortage, can cause such an operations dilemma (called "mistakes"), then JIT is no longer the best approach to inventory management.   Suppose, for example, an organization allows employees to shut down a production line to avoid a quality problem.   But in shutting down the production line the total costs of the down line (now and in the near future) are greater than the quality problem.   In such a situation, the JIT principle of empowerment to the employee is not the most economically wise policy.   Only if learning takes place, will correcting the problem now and in the future generate possible long-term benefits.   And in finding and eliminating the quality problem can JIT hope to pay for its self and justify the JIT principles as the most successful approach to management. Ocana and Zemel (1996) felt that waiting for a mistake is not proactive enough to make JIT a successful operations strategy in some production environments. They developed and presented a series of proofs that supported their anti-JIT hypothesis.

What can we conclude from the theoretical work of Ocana and Zemel (1996)?   First, all the JIT training in the world is useless unless learning takes place.   Learning is what will prevent costly mistakes that would render a JIT operation more costly than some other management inventory or production planning approach.   Specifically, HRM managers should only embark on training programs that have proven learning capabilities.   Second, under JIT, learning may only occur if a mistake is found.   In other words, if the system is working well, don't fix it.   This is counter to the "continuous improvement" component that is also a part of JIT.   Third, Ocana and Zemel (1996) specifically recommended that performance criteria used to judge JIT system performance, should be changed periodically to insure that change is continual.   Finally, they also pointed out that firms sometimes adopt characteristics of JIT operations without adopting other JIT characteristics that inherently lead to sequential improvement in the operating systems.

JIT AND EMPOWERMENT

Empowering employees is one of the basic principles on which JIT rests. Empowerment starts with respect for employees and unions (see Heaton, 1998). Companies can empower the individual employee and empower teams of individuals.   Empowerment puts the opportunity to improve operations in the hands of those workers who are best equipped to suggest and implement improvements.

Most empowerment programs begin with employee education in areas of knowledge like JIT.   But empowerment means an education that will never end. The principle of continuous improvement should be applied to the employee. Clear and achievable educational goals must be set for individuals and teams. The nurturing necessary to sustain the educational efforts comes from respecting the efforts of the individual or team.   As employees learn more, they become a greater asset to the organization, and as a result, achieve a type of guaranteed employment or job security necessary to motivate employees to suggest ideas that might otherwise cost them their own jobs.

JIT research (see Mullarkey, Jackson and Parker, 1995) examining the impact on employees of empowerment programs reveals a number of benefits to the employees and employers. The psychological well-being, job satisfaction, job-related anxiety, and job-related depression were all found to be improved with the application of JIT principles and their related empowerment. This research was particularly interesting in that it showed that group climate was very positively improved with empowerment. The study showed that co-worker support and group cohesiveness measures related to the "bringing together" of team's members into production cells had a positive impact on employees. Unlike prior research (see Klein, 1991) that had suggested the JIT pushed and imposed unrealistic demands on employees, Mullarkey, Jackson, and Parker's (1995) empirically based study disproved the critics of JIT and showed that empowering employees by letting them assume the maximum control over managing quality and production scheduling, resulted in a successful implementation of JIT and its positive benefits for the organization.

What the research of Mullarkey, Jackson, and Parker (1995) means to practicing managers is simply this: Organizations that take a HRM developmental approach to introducing JIT will reap many benefits. A participatory approach that focuses on the human resources by providing the multiskilled training and JIT education before reducing inventory will result in a better social climate for acceptance and use of JIT principles. Conversely, organizations that just start cutting inventory and provide the JIT education when a problem arises, will not maximize the benefits of JIT in either the short-term or long-term.

On the other hand, empowerment as Heaton (1998) pointed out, is not complete freedom to do as much or as little as one wants. In JIT it should be the shared joining of labor and management to achieve the common goals of mutual respect for the contribution of each other, and the elimination of waste where ever it may be.

JIT AND LABOR RELATIONS

Recent research on issues in labor relations, including "forms of employment," "seniority," and "trade unions" reveals a number of differences between JIT Japanese and JIT Swedish firm practices (see Dahlem, Ericsson, and Fujii, 1995). These country differences can have generalizable advantages for all firms.

In the various forms of employment offered to employees, the research revealed that Japanese firms using JIT had a greater variety in the forms of employment offered to workers than the Swedish JIT counterparts. These Japanese firms used full-time (called "core") employees, part-time employees, and temporary employees in much the same way that U.S. firms now use the same three forms of employment. In addition, the Japanese JIT firms also had a category of employment called the, "old boys" that consisted of bringing back retired employees to do easier tasks. Another category of employment used was "leased employees" (i.e., leased from other companies) to perform routine or monotonous tasks.

Seniority was fully rewarded in Japanese JIT firms when compared to the Swedish JIT firms. The results of the study by Dahlem, Ericsson, and Fujii (1995) showed that comparable seniority increases in wage levels were more than four times those of Swedish JIT firms.

Trade unions are differently organized by country as well. In the Japanese JIT firms, one trade union covered an average of 86 percent of all employees. In the Swedish JIT firms, different unions represented different groups of employees. In the production area, different unions were used to cover the blue-collar workers, foreman, officials, and engineers. When asked what the pressing trade union issues were, the Japanese JIT employees cited consensus, working hours, wages/bonuses, and working conditions (i.e., work environment). The Swedish JIT trade union employees cited similar issues as the Japanese, except they added job security and equality in employment between the sexes.

What the Dahlem, Ericsson, and Fujii (1995) study reveals to practicing managers is that there are still differences in the forms of employment given workers in JIT operations, differences in the way labor relations are handled, and issues like seniority and trade unions are dealt with in Eastern and Western JIT operations. Does the difference make one approach wrong? No, but opportunities exist by taking advantage of what may work in other countries like the United States. For example, increased use of retired employees to fill shortages in JIT production cells may provide a more knowledgeable, and therefore, more productive, employee than just using temporary help to fill in for absentee personnel.

JIT AND LAYOUT DESIGN

Layout design is usually performed by industrial engineers. While most industrial engineers are trained in JIT design principles. Unfortunately, they as group do not always consider the JIT HRM component in the layout design. Considering the human side of JIT layout design is an advanced topic rarely included in JIT training programs of design. Current areas of layout design considerations include the impact of HRM in the design of production cells, the visualization provided by the layout to employees, and work performance measures used in JIT layout systems.

The design of a JIT production cell usually consists of the typical U-shaped or C-shaped cell design. This layout design has been shown to facilitate more effective operations and rapid problem solving by team members while also improving morale by the perception of fair team load sharing (see Mullarkey, Jackson and Parker, 1995). Indeed, research has shown that the U-shaped JIT cells result in positive changes in group climate by simply placing people in closer proximity. To ensure the maximum benefit from cell layout design, several JIT HRM design factors should be considered. These design factors include:

1. *Analysis of current tasks and skills.* A JIT production cell will require a wide range of skills to meet the multi-task functions placed on employees. Questions to answer here include: Does the employee have the necessary skills to perform the tasks, operate the equipment, and possess the multi-function behaviors required in the production cell? Have the employees been trained in JIT reporting methods (e.g., quality control reports, statistical control charting, etc.)? Is the employee familiar with the processes and the procedures they must follow in performing their cell tasks?

2. *Analysis the fit between the abilities of the employee and the JIT tasks expected in the production cell.* Matching up the right person with the abilities to do the job is essential in the sometime pressure oriented JIT operation. Questions to answer here include: Can the employee move fast enough between workstations in the cell to achieve the necessary flexibility in work flow and mixed model scheduling? Does the employee possess the manual dexterity to accomplish the required JIT tasks? Does the employee possess the mental capabilities to perform the diverse tasks expected in the cell?

3. *Analysis the potential constraints on employee and JIT cell performance.* In any production plant there are work rules to guide employees and managers in the performance of their tasks. Questions to answer here include: Are there work rules or union restrictions that might motivate noncompliance from employees in basic JIT task responsibilities? Are the work stations in the cell designed to minimize employee effort to accomplish their tasks? Are the performance measures used to gauge productivity and employee work performance accurate and useful?

4. *Develop a strategy of cooperation that will resolve any problems found and reported in the design layout.* Change toward a JIT operation will result in the same kind of problems that any change in production methods and processes entail. Even though JIT principles are based on cooperation, there is a special need to make sure that layout issues, particularly the HRM ones, can be resolved quickly. Questions to answer here include: Are there an easy means of communication between the employees and management to correct production cell design problems? Are there an easy means to communication between managers and design engineers? Are there means of communication and problem resolution between management and labor to resolve layout problems?

The issue of visualization in JIT layout design is one of the basic principles required in JIT operations (see Moden, 1998). In JIT firms there is two types of visualization: immediate and long-range (see Dahlem, Ericsson, and Fujii, 1995). *Immediate visualization* utilizes technologies, like *Andon boards* (i.e., warning lights placed on boards by specific production criteria) to prominently give the current status of production line activities. These visual displays are used to allow employees to see how well their performance is matching up against the production quotes or expectations in quality. *Long-range visualization* involves the display of statistics, diagrams, and matrices to depict longer-term production standards and on-going production line activities, like downtime expectations and actual downtime behavior. While research has shown that not all organizations use visualization methods (see Dahlem, Ericsson and Fujii, 1995), they have been viewed for many years as a basic JIT motivational tools. Western organizations have been viewed as lacking visualization criteria such as productivity, suggestion and idea counts, and multiskill accomplishments of personnel in their visualization layout design.

Another layout design consideration includes JIT employee work performance measures. *JIT employee work performance measures* include such items as productivity measures, quality measures, inventory measures, lead-time measures, preventive maintenance, scheduling performance, and other cost measures. When a new layout design is undertaken, the work efforts performed by employees using the new layout will change. This change can be very substantial when converting to a JIT layout design. These changes, and the new expectations for employees in a JIT operation require changes in the performance measures taken and used on the shop floor. Research by Marsh and Meredith (1998) have suggested the implementation of JIT production cells has the tendency to reduce the use of productivity and scheduling measures, but increase the use of quality, lead time, inventory, preventive maintenance and other cost measures for measuring operating performance. While these research observations suggest the current JIT practices, they may also portend a missed opportunity for JIT managers.

SUMMARY

This chapter presented a series of HRM issues and discussed how they can impact JIT management. These issues included: teamship, learning, empowerment, labor relations, and layout design. For each of the HRM issues raised, current research were presented to explain ideas and strategies to overcome problems and maximize JIT performance.

HRM is a still a functional area in most business organizations. In the context of a JIT organization, HRM holds a particularly critical role: acquiring personnel who can specifically or best help JIT become a reality. As all practicing managers know, the role of employees, more than technology, processes or systems, determines the successfulness of JIT in all organizations. When HRM departments undertake their tasks of personnel recruiting, selecting, and training JIT employees requires an understanding of personnel characteristics that best support JIT activities. This chapter has focused on issues that recent advances in research suggest are the best approaches in making JIT successful.

This chapter completes the basic chapters of this book. Like the dynamic behavior of human resources at work, nothing ever stays the same. If a person is taught how to perform a task, we know from learning curves and basic human nature, change will be introduced to alter the task. Like the human resources that make up JIT, it is a dynamic force that seeks, motivates, and even demands change take place. This change can be seen not only in what takes place on a JIT shop floor, but in the literature that reports on JIT. The diversity of literature used in the content of the previous chapters is a current, but a small component of the total wealth of knowledge available on the subject of JIT. In the appendices of this book can be found a fairly complete listing of both recent JIT research books (Appendix I) and JIT journal articles (Appendix II). While the content of this book has been focused on the current advances in JIT research, the historical perspective that can be found in these appendices can add greater substance and greater purpose to the ideas presented in this book.

REFERENCES

Abraham, S. E., and Spencer, M. S. "The Legal Limitations to Self-Directed Work Teams in Production Planning and Control," *Production and Inventory Management Journal*, 39 (1998), pp. 41-45.

Adair-Heely, C. *The Human Side of Just-In-Time: How to Make the Techniques Really Work*. New York: American Management Association, 1991.

Conti, R. F. "Variable Manning JIT: An Innovative Answer to Team Absenteeism," *Production and Inventory Management Journal*, 37 (1996), pp. 24-27.

Dahlem, P., Ericsson, J., and Fujii, H. "Labor Stability and Flexibility-Conditions to Reach Just-In-Time," *International Journal of Operations and Production Management*, 15 (1995), pp. 26-44.

Deshpande, S. P., and Golhar, D. Y. "HRM Practices in Unionized and Nonunionized Canadian JIT Manufacturing," *Production and Inventory Management Journal*, 36 (1995), pp. 15-19.

Deshpande, S. P., and Golhar, D. Y. "HRM Practices of JIT Firms in Canada," *Production Planning and Control*, 7 (1996), pp. 79-85.

Deshpande, S. P., Golhar, D.Y., and Stamm, C. L. "Human Resource Management in the Just-In-Time Environment," *Production Planning and Control*, 5 (1994), pp. 372-380.

Gould, W. B. "Employee Participation and Labor Policy: Why Team Act Should be Defeated and the National Labor Relations Act Amended," *Creighton Law Review*, 30 (1996), p. 3.

Heaton, W. E. "The Secret Strategy," *Production and Inventory Management Journal*, 39 (1998), pp. 78-81.

Heeley, C. B. *The Human Side of Just-In-Time*. New York: American Management Association, 1991.

Klein, J. A. "A Re-examination of Autonomy in Light of New Manufacturing Practices," *Human Relations*, 44 (1991), pp. 21-38.

Luthans, F. *Organizational Behavior*, 8th ed. Boston: Irwin and McGraw-Hill, 1998.

Marsh, R. F., and Meredith, J. R. "Changes in Performance Measures on the Factory Floor," *Production and Inventory Management Journal*, 39 (1998), pp. 36-40.

Moden, Y. *Toyota Production System: An Integrated Approach to Just-In-Time*, 3 rd ed., New York: Engineering and Management Press, 1998.

Mullarkey, S., Jackson, P. R., and Parker, S. K. "Employee Reactions to JIT Manufacturing Practices: A Two-phase Investigation," *International Journal of Operations and Production Management*, 15 (1995), pp. 62-80.

Ocana, C. and Zemel, E. "Learning from Mistakes: A Note on Just-In-Time Systems," *Operations Research*, 44 (1996), pp. 206-214.

Price, K. "Tearing Down the Walls: The Need for Revision of NLRA Sec.8a2 to Permit Management-Labor Participation Committees to Function in the Workplace," *University of Cincinnati Law Review*, 63 (1995), p. 1379.

Yasin, M. M., Small, M., and Wafa, M. A. "An Empirical Investigation of JIT Effectiveness: An Organizational Perspective," *Omega*, 25 (1997), pp. 461-471.

Yasin, M. M., and Wafa, M. A. "An Empirical Examination of Factors Influencing JIT Success," *International Journal of Operations and Production Management*, 16 (1996), pp. 19-27.

APPENDIX I

SOURCES OF ADDITIONAL JIT BOOK INFORMATION

This appendix presents a comprehensive bibliography of JIT and JIT-related books. The purpose of this appendix is to provide additional sources of JIT information. These books were identified using a variety of business, engineering, and economics databases, as well as the usual manual effort of library searches. These books range from those published in 1981 to 1998 are listed alphabetically by author.

To help users identify useful books, each has been coded with one or more letters in brackets after the period at the end of the citation. The codes can be used to quickly identify whether the book is focused on a narrow topic or cover a wide range of JIT topics.

Code Letter	Topic
A:	Accounting Methods
B:	Introductory Concepts, Application of Principles, Broad Range of Ideas
C:	Case Studies
D:	Design Concepts
H:	Human Resource Management
K:	Kanban Analysis, Scheduling
P:	Purchasing, Supplier Relations, Supply chain, Logistics, Material Handling
M:	Mathematical Methods, Quantitative Procedures
Q:	Quality Management

JIT BOOK BIBLIOGRAPHY LISTING

Adair-Heeley, C. *The Human Side of Just-in-time : How to Make the Techniques Really Work*. New York: American Management Association, 1991. [H, B]

Ansari, A., and Modarress, B. *Just-in-Time Purchasing*. London: Collier Macmillan Publishers, 1990. [P]

Bare, L. *The Self-instructional Route to Statistical Process Control and Just-in-time Manufacturing*. Milwaukee, WI: ASQC Quality Press, 1991. [Q, M]

Barefield, R. M., and Young, M. S. *Internal Auditing in a Just-in-time Manufacturing Environment*. Altamonte Springs, FL: Institute of Internal Auditors Research Foundation, 1988. [A, M]

Bicheno, J. *Implementing JIT: How to Cut Out Waste and Delay in Any Manufacturing Operation*. Bedford, England: IFS Publications, 1991. [B]

Black, J. T. *The Design of the Factory With a Future*. New York: McGraw-Hill, 1991. [D]

Blackburn, J. D. *Time-based Competition: The Next Battleground in American Manufacturing*. Homewood, IL: Business One Irwin, 1991. [B]

Bragg, S. M. *Just-in-time Accounting: How to Decrease Costs and Increase Efficiency*. New York: Wiley & Sons, 1996. [A, M]

Burnham, J. H. *Just-in-Time in a Major Process Industry: Case Studies of JIT Implementation at Alcoa*. Milwaukee, WI: American Production & Inventory Control Society, Inc., 1986. [B, C]

Cheng, T. C. E., and Podolsky, S. *Just-in-time Manufacturing: An Introduction*, 2 nd ed., London: Chapman & Hall, 1996. [B]

Cammarano, J. J. *Lessons to Be Learned Just in Time*. New York: Engineering Management Press, 1997. [B]

Crawford, K. M., Cox, J. F., and Blackstone, J. F. *Performance Measurement Systems and the JIT Philosophy: Principles and Cases*. Milwaukee, WI: American Production & Inventory Control Society, Inc., 1988. [M, C]

Dear, A. *Working Towards Just-in-time*. New York: Van Nostrand Reinhold, 1988. [B]

Duin, S., Barlow, W., Kapoor, V., Bullock, R. and Schlichting, W. *Case Studies of Just-in-Time Implementation at Westinghouse & IBM*. American Production & Inventory Control Society Staff (Editor), Milwaukee, WI: American Production & Inventory Control Society, Inc., 1986. [C]

Duncan, W. L. *Just-In-Time in American Manufacturing*. Dearborn, MI: Society Manufacturing Engineers, 1988. [B]

Fisher, D. *The Just-in-time Self Test: Success Through Assessment and Implementation*. Chicago: Irwin Professional Publications, 1995. [B]

Goddard, W. E. *Just-in-time: Surviving by Breaking Tradition.* Essex Junction, VT: Oliver Wight Ltd. Publications, 1986. [B]

Graham, I. *Just-In-Time Management of Manufacturing.* Letchworth, UK: Technical Communications, 1988. [B]

Grieco, P. L. *World Class: Measuring Its Achievement.* New York: P T Publications, Inc, 1990. [M]

Grieco, P. L., Gozzo, M. W., and Claunch, J. *Just-in-Time Purchasing: In Pursuit of Excellence.* New York: P. T. Publications, Inc., 1988. [P]

Hall, R. W. *Attaining Manufacturing Excellence: Just-in-Time, Total Quality, Total People Involvement.* Homewood, IL: Dow Jones-Irwin, 1987. [Q, H]

Hall, R. W., *Implementation of Zero Inventory: Just in Time.* Milwaukee, WI: American Production & Inventory Control Society, Inc., 1986. [B]

Harding, M. *Profitable Purchasing: An Implementation Handbook for Just-in-Time.* New York: Industrial Press, Inc. [P]

Harrison, A. *Just-in-time Manufacturing in Perspective.* London: Prentice Hall International, 1992. [B]

Hay, E. J. *The Just-in-Time Breakthrough: Implementing the New Manufacturing Basics.* New York: Wiley and Sons, Inc., 1988. [B]

Hernandez, A. *Just-in-Time Manufacturing: A Practical Approach.* Englewood Cliffs, NJ: Prentice-Hall, 1989. [B]

Hernández del Campo, A. *Just-in-Time Quality: A Practical Approach,* 2 nd ed., Englewood Cliffs, NJ: PTR Prentice Hall, 1993. [Q]

Hirano, H. *JIT Factory Revolution: A Pictorial Guide to Factory Design of the Future.* Cambridge, MA: Productivity Press, 1988. [B, D]

Holl, U., and Trevor, M. *Just in Time Systems & EURO-Japanese Industrial Collaboration.* New York: Westview Press, 1998. [B]

Ichuro, M. *Shift to JIT: How People Make.* New York: Productivity Press, Inc., 1994. [B]

Japan Manufacturing Association, *Kanban/Just-In-Time at Toyota.* Cambridge, MA: Productivity Press, 1989. [K]

Kaynak, H. *Total Quality Management and Just-in-Time Purchasing: Their Effects on Performance of Firms Operating in the U.S.* New York: Garland Publishers, 1997. [B]

Keane, P. T., and King, J. T. *Failing in the Factory: A Shop Floor Perspective on Correcting America's Misunderstanding & Misuse of Just-in-Time.* New York: Brown House Communications, 1991. [B]

Kliem, R.L., and Ludin, I. S. *Just-in-Time Systems for Computing Environments.* Westport, CT: Quorum Books, 1994. [B, D]

Ling, R. C., and Goddard, W. E. *Orchestrating Success: Improve Control of the Business with Sales & Operations Planning*. Essex Junction, VT: Oliver Wight Ltd. Publications, 1988. [B]

Louis, R. S. *Integrating Kanban with MRPII: Automating a Pull System for Enhanced JIT Inventory Management*. Portland, OR: Productivity Press, 1997. [K]

Lu, D. J., and Kyokai, N. N. *Kanban Just-in-time at Toyota: Management Begins at the Workplace*. Portland, OR: Productivity Press, 1986. [B, K]

Lubben, R. T. *Just-in-time Manufacturing: An Aggressive Manufacturing Strategy*. New York: McGraw-Hill, 1988. [B]

Maskell, B. H. *Software & the Agile Manufacturer: Computer Systems & World Class Manufacturing*. Portland, OR: Productivity Press, Inc., 1990. [D]

McNair, C. J., Mosconi, W., and Norris, T. *Beyond the Bottom Line: Measuring World Class Performance*. Homewood, IL: Dow Jones-Irwin, 1989. [B, M]

Merli, G. *Total Manufacturing Management: Production Organization for the 1990s*. Cambridge, MA: Productivity Press, 1990. [B]

Monden, Y. *Toyota Production System: An Integrated Approach to Just-in-Time*, 3 rd ed., New York: Engineering & Management Press, 1998. [B]

O'Neal, C. R, and Bertrand, K. *Developing a Winning J.I.T. Marketing Strategy: The Industrial Marketer's Guide*. Englewood Cliffs, NJ: Prentice Hall, 1990. [B]

Petroff, J.N. *Handbook of MRP II and JIT: Strategies for Total Manufacturing Control*. Englewood Cliffs, NJ: Prentice-Hall, 1993. [K]

Productivity Press Development Team Staff, *Just in Time for Operators*. Cambridge, MA: Productivity Press, 1998. [B, H]

Rutherford, R. D. *Just-in-time: Immediate Help for the Time-pressured*. New York: Wiley & Sons, Inc., 1981. [B]

Sandras, W. A. *Just-in-Time: Making it Happen: Unleashing the Power of Continuous Improvement*. Essex Junction, VT: Oliver Wight Ltd. Publications, 1989. [B]

Schniederjans, M. J. *Topics in Just-in-time Management*. Boston: Allyn and Bacon, 1993. [B]

Schonberger, R. J. *Just-in-time: A Comparison of Japanese and American Manufacturing*. Norcross, GA: Industrial Engineering and Management Press, 1984. [B]

Schonberger, R. J. *World Class Manufacturing: The Lessons of Simplicity Applied*. New York : Free Press, 1986. [B]

Schonberger, R. J. *World Class Manufacturing: The Next Decade: Building Power, Strength, & Value*. New York: The Free Press, 1996. [B]

Schonberger, R. J. *World Class Manufacturing Casebook: Implementing JIT and TQC*. New York: The Free Press, 1995. [C, Q, B]

Schorr, J. E. *Purchasing in the 21 st Century: A Guide to State of Art Techniques and Strategies*. New York: John Wiley & Sons, Inc., 1998. [P]

Stasey, R., and McNair, C. J. *Crossroads: A JIT Success Story*. Homewood, IL: Dow Jones-Irwin, 1990. [C]

Trevor, M., and Holl, U. *Just-in-time Systems and Euro-Japanese Industrial Collaboration* ed. F. Main, and C. Verlag, Boulder, CO: Westview Press, 1988. [B]

Vaessen, W. *The Economics of Speed: Assessing the Financial Impact of the Just-in-time Concept in the Chemical-Pharmaceutical Industry*. New York: Bern, 1992. [M,C]

Voss, C. *Just-in-Time Manufacture*. Bedford, England: IFS Publications, 1987. [B]

Voss, C., and Clutterbuck, D. *Just-in-Time: A Global Status Report*. New York: Springer-Verlag, 1989. [B]

Wantuck, K. *Just-in-Time for America: A Common Sense Production Strategy*. Milwaukee, WI: K W A Media, 1989. [B]

Wheatley, M. *Understanding Just-in-Time in a Week*. London: Hodder & Stoughton, 1992. [B]

Womack, J. P., Jones, D. T., and Roos, D. *The Machine That Changed the World*. New York: Collier Macmillan, 1990. [B]

Wu, A. *International Conferences Accounting Papers Collection: Papers on Earnings Forecasts, Business Combinations, JIT, and Job Stress for Accounting Employees*. Taipei, Taiwan: Cheng Yih Culture Enterprise Co, 1993. [A, M, H]

Appendix II

Sources of Additional JIT Journal Article Information

This appendix presents one of the most comprehensive bibliographies of JIT *Journal* articles to date. The purpose of this appendix is to provide additional sources of JIT information. These articles were collected using a variety of business, engineering, and economics databases. The articles range from those published in 1982 to 1999. Since the collection effort began early in 1998, the natural lateness of the publication of journal articles (i.e., lead time from when they are dated from when they are actually published and sent to libraries) may have prevented a comprehensive location of the actual publication from both of the years 1998 or 1999. Also, some of the older JIT articles in the early 1980s may not have been listed in the databases used in this bibliography.

Why did we select the years of 1982 to 1999? Contained in the bibliography are numerous JIT prior article bibliographies that have substantially covered the articles prior to 1982. We also feel that this period represents a more homogeneous group of research than the exploratory years prior to 1982.

To help users identify useful articles, each has been coded with one or more letters after the period at the end of the citation. The codes can be used to quickly identify the main topic of the article as they relate to the topical subjects in this book. If no code is given, the article should be considered a miscellaneous type paper with general applicability to a wide range of subjects, too broad to classify using a couple of codes.

Code Letter	Topic
C:	Introductory Concepts, Application of Principles [Chapter 1]
H:	Human Research Management [Chapter 8]
J:	Economic Justification [Chapter 2]
K:	Kanban Analysis, Scheduling [Chapters 6 and 7]
P:	Purchasing, Supplier Relations, Supply-Chain, Logistics, Material Handling [Chapters 4 and 5]
S:	Simulation [Chapter 3]

As mentioned in book, there are many other topics than those covered in chapter in this book that are very relevant to JIT journal article research. The following additional codes of topical areas in JIT can also be used to help classify the articles:

Code Letter	Topic
E:	Empirical Research
I:	International Study
M:	Maintenance
N:	New Concept, New Methodology
Q:	Quality Management

JIT ARTICLE BIBLIOGRAPHY LISTING

Abdou, G., and Dutta, S. P. "A Systematic Simulation Approach for the Design of JIT Manufacturing Systems," *Journal of Operations Management*, 11 (1993), pp. 225-238. [S]

Abdul-Nour, G. "On Some Factors Affecting the Just-In-Time Production System Output Variability: A Simulation Study Using Taguchi Techniques," *Computers and Industrial Engineering*, 25 (1993), pp. 461-464. [S]

Abdul-Nour, G., Dudek, R. A., and Smith, M. L. "Effect of Maintenance Policies on the Just-In-Time Production System," *International Journal of Production Research*, 33 (1995), pp. 565-583. [M]

Abdul-Nour, G., Serge Lamb, S., and Drolet, J. "Adaptation of JIT Philosophy and Kanban Technique to Small-sized Manufacturing Firm: A Project Management Approach," *Computers & Industrial Engineering*, 35 (1998), pp. 419-423. [K]

Abraham, Steven E., and Spencer, M. S. "The Legal Limitations to Self-directed Work Teams in Production Planning and Control," *Production & Inventory Management Journal*, 39 (1998), pp. 41-46. [H]

Abraham, Y., Holt, T., and Kathawala, Y. "Just-In-Time: Supplier-Side Strategic Implications, *Industrial Management and Data Systems*, 3 (1990), pp. 12-17. [P]

Adachi, T., Enkawa, T., and Shih, L. "A Concurrent Engineering Methodology Using Analogies to Just-In-Time Concepts," *International Journal of Production Research*, 33 (1995), pp. 587-605. [N]

Aderohunmu, R., Mobolurin, A., and Bryson, N. "Joint Vendor-Buyer Policy in JIT Manufacturing," *Journal of the Operational Research Society*, 46 (1995), pp. 375-385. [P]

Aderohunmu, R, Mobulurin, A., and Bryson, N. "Joint Vendor-Buyer Policy in JIT Manufacturing: Response to Hofmann's Comments," *OR: The Journal of the Operational Research Society*, 48 (1997), pp. 547-555. [P]

Agrawal, A., Harhalakis, G., Ioannis, M., and Nagi, R. " 'Just-In-Time' Production of Large Assemblies," *IIE Transactions*, 28 (1996), pp. 653-567. [C]

Aggrawal, S. "MRP, JIT, FMS? Making Sense of Production Operations Systems," *Harvard Business Review*, 63 (1985), pp. 8-16. [C]

Ahmed, N. U., Tunc, N. A., and Montagno, R. V. "A Comparative Study of US Manufacturing Firms at Various Stages of Just In Time Manufacturing," *International Journal of Production Research*, 29 (1991), pp. 757-802. [E]

Aigbedo, H., and Monden, Y. "A Simulation Analysis for Two-Level Sequence-Scheduling for Just-In-Time (JIT) Mixed-Model Assembly Lines," *International Journal of Production Research*, 34 (1996), pp. 3107-3124. [S, K]

Aigbedo, H., and Monden, Y. "A Parametric Procedure for Multicriterion Sequence Scheduling for Just-In-Time Mixed-Model Assembly Lines," *International Journal of Production Research*, 35 (1997), pp. 2543-2062. [K]

Akinc, U. "Selecting a Set of Vendors in a Manufacturing Environment," *Journal of Operations Management*, 11 (1993), pp. 107-122. [P]

Al-Asseri, I., and Hariga, M. "A Simple Lot-Scheduling Model for Just-In-Time Manufacturing," *International Journal of Production Research*, 33 (1995), pp. 3143-3161. [K]

Albino, V., Carella, G., and Okagbaa, O. G. "Maintenance Policies in Just In Time Manufacturing Lines," *International Journal of Production Research*, 30 (1992), pp. 369-382.. [M]

Albino, V., Dassisti M., and Okogbaa, G. O. "Approximation Approach for the Performance Analysis of Production Lines Under a Kanban Discipline," *International Journal of Production Economics*, 40 (1995), pp. 197-207. [K]

Albino, V., and Garavelli, A. C. "A Methodology for the Vulnerability Analysis of Just-In-Time Production Systems," *International Journal of Production Economics*, 41 (1995), pp. 71-80. [N]

Alles, M., Datar, S. M., and Lambert, R. A. "Moral Hazard and Management Control in Just-In-Time Settings," *Accounting Research*, 33 (1995), pp. 177-204. [H]

Alnestig, P., and Segerstedt, A. "Product Costing in Ten Swedish Manufacturing Companies," *International Journal of Production Economics*, 46 (1996), pp. 441-467. [I]

Alonso, R. L., and Frasier, C. W. "JIT Hits Home: A Case Study in Reducing Management Delays," *Sloan Management Review*, 32 (1991), pp. 59-67. [C]

Anderson, D., and Quinn, R. J. "The Role of Transportation in Long Supply Line Just In Time Logistics Channels," *Journal of Business Logistics*, 7(1986), pp. 68-88. [P]

Andijani, A, Atkinson, Helen and Baldwin, Tom "A Multi-Criterion Approach for Kanban Allocations," *Omega*, 26 (1998), pp. 483-492. [K]

Ansari, A. "Strategies for the Implementation of JIT Purchasing," *International Journal of Physical Distribution and Materials Management*, 29 (1991), pp. 1-15. [P, C]

Ansari, A., and Hechel, J. "JIT Purchasing: Impact of Freight and Inventory Costs," *Journal of Purchasing and Material Management*, 23 (1987), pp. 24-28. [P]

Ansari, A., and Modarress, B. "JIT Purchasing as a Quality and Productivity Center," *International Journal of Production Research*, 26 (1988), pp. 19-26. [P, Q]

Ansari, A., and Modarress, B. "The Potential Benefits of Just In Time Purchasing for US Manufacturing," *Production and Inventory Management Journal*, 28 (1987), pp. 30-35. [C]

Ansari, A., and Modarress, B. "Transportation Systems for Meeting Just-In-Time Materials Delivery Requirements," *Production Planning and Control*, 2 (1991), pp. 273-279. [N]

Ansari, A., and Modarress, B. "Wireless Kanban," *Production and Inventory Management Journal*, 36 (1995), pp. 60-65. [K]

Antonio, E. J., and Egbelu, P. J. "Design of Synchronous Manufacturing System with Just In Time Production," *Computers and Industrial Engineering*, 21 (1991), pp. 553-563. [C]

Anwar, M. F., and Nagi, R. "Integrated Lot Sizing and Scheduling for Just-In-Time Production of Complex Assemblies with Finite Setups," *International Journal of Production Research*, 35 (1997), pp. 1447-1470. [K]

Anyane-Ntow, K. "Just-In-Time Manufacturing Systems and Inventory Reported in Financial Statements: A Cross-National Comparison of Manufacturing Firms," *International Journal of Accounting*, 26 (1991), pp. 277-285. [J]

Arogyaswamy, B., and Simmons, R. P. "Thriving on Interdependencies: The Key to JIT Implementation," *Production and Inventory Management Journal*, 32 (1991), pp. 56-60. [C]

Askin , R. G., Mitwasi, G. M., and Goldberg, J. B. "Determining the Number of Kanbans in Multi-Item Just-In-Time Systems," *IIE Transactions*, 25 (1993), pp. 89-98. [K]

Aull-Hyde, R., Gempesaw II, C. M., and Sundaresan, I. "Procurement Policies in the U. S. Broiler Industry: Shall We Call them JIT?," *Production and Inventory Management Journal*, 35 (1994), pp. 11-16. [P]

Axsater, S., and Rosling, K. "Ranking of Generalised Multi-stage KANBAN Policies," *European Journal of Operational Research*, 113 (1999), pp. 560-568. [K]

Aytug, Haldun, and Dogan, A. "A Framework and a Simulation Generator for Kanban-controlled Manufacturing Systems," *Computers & Industrial Engineering*, 34 (1998), pp. 337-361. [S, K]

Bagchi, P. K. "Management of Materials Under Just In Time Inventory System: A New Look," *Journal of Business Logistics*, 9 (1988), pp. 63-73. [P]

Bagchi, P. K., Raghunathan, T. S., and Bardi, E. J. "The Implications of Just In Time Inventory Policies on Carrier Selection," *Logistics and Transportation Review*, 23 (1987), pp. 373-384. [P]

Baker, R. C., Chang, R. E., and Chang, I. C. "Switching Rules for JIT Purchasing," *Production and Inventory Management*, 35 (1994), pp. 13-17. [P]

Balakrishnan, Nagraj, Kanet, John J. and Sridharan, Sri V. "Early/tardy Scheduling with Sequence Dependent Setups on Uniform Parallel Machines," *Computers & Operations Research*, 26 (1999), pp. 127-137. [K]

Bandyopadhyay, J., "Product Design to Facilitate JIT Production," *Production and Inventory Management*, 31 (1990), pp. 71-76. [C]

Banerjee, P., and Armouti, H. A. "JIT Approach to Integrating Production Order Scheduling and Production Activity Control," *Computer-Integrated Manufacturing Systems*, 5 (1992), pp. 283-290. [K]

Banerjee, A., and Kim, S. L. "An Integrated JIT Inventory Model," *International Journal of Operations and Production Management*, 15 (1995), pp. 237-244. [C]

Banerjee, S., and Golhar, D.Y. "EDI Implementation in JIT and Non-JIT Manufacturing Firms: A Comparative Study," *International Journal of Operations and Production Management*, 13 (1993), pp. 25-37. [P, E]

Banerjee, S., and Golhar, D. Y. "EDI Implementation: A Comparative Study of JIT and Non-JIT Manufacturing Firms," *International Journal of Physical Distribution and Logistics Management*, 23 (1993), pp. 22-31. [P, E]

Barker, B. "Value-Adding Performance Measurement: A Time-Based Approach," *International Journal of Operations and Production Management*, 13(1993), pp. 33-40. [J]

Bartezzaghi, E., and Turco, F. "Impact of JIT Production System Performance: An Analytical Framework," *International Journal of Operations and Production Management*, 9 (1989), pp. 40-62. [C]

Bartezzaghi, E., Turco, F., and Spina, G. "The Impact of the Just-In-Time Approach on Production Performance: A Survey of Italian Industry," *International Journal of Operations and Production Management*, 12 (1992), pp. 5-17. [I]

Bautista, J., Companys, R., and Corominas, A. "A Note on the Relation Between the Product Rate Variation (PRV) Problem and the Apportionment Problem," *Journal of the Operational Research Society*, 47 (1996), pp. 1410-1414. [J]

Baykoc, O. F., and Erol, S. "Simulation Modeling and Analysis of a JIT Production System," *International Journal of Production Economics*, 55 (1998), pp. 203-207. [S]

Bechte, W. "Load-Oriented Manufacturing Control Just-In-Time Production for Job Shops," *Production Planning and Control*, 5 (1994), pp. 292-307. [K]

Bedwell-Brace, S., and Inman, R. A. "Simulating JIT Manufacturing: Prospects and Pitfalls," *Industrial Management and Data Systems*, 91 (1991), pp. 3-5. [S]

Belt, B. "MRP and Kanban - A Possible Synergy," *Production and Inventory Management Journal*, 28 (1987), pp. 71-79. [K]

Ben-Daya, M., and Raouf, A. "Inventory Models Involving Lead Time as a Decision Variable," *Journal of the Operational Research Society*, 45 (1994), pp. 579-582. [J]

Benton, W. C., and Shin, H. "Manufacturing Planning and Control: The Evolution of MRP and JIT Integration," *European Journal of Operational Research*, 110 (1998), pp. 411-416. [K]

Berkley, Blair J. "A Decomposition Approximation for Periodic Kanban-Controlled Flow Shops," *Decision Sciences*, 23 (1992), pp. 291-311. [K]

Berkley, Blair J., and Kiran, Ali S. "A Simulation Study of Sequencing Rules in a Kanban-Controlled Flow Shop," *Decision Sciences*, 22 (1991), pp. 559-572. [S, K]

Berkley, Blair J. "Analysis and Approximation of a JIT Production Line: A Comment," *Decision Sciences*, 21 (1990), pp. 660-669. [S]

Bhask, Sita "Simulation Analysis of a Manufacturing Supply Chain," *Decision Sciences*, 29 (1998), pp. 633-42. [P, S]

Billesbach, T. "A Study of the Implementation of Just-In-Time in the United States," *Production and Inventory Management Journal*, 32 (1991), pp. 1-4. [C]

Billesbach, T. "Simplified Flow Control Using Kanban Signals," *Production and Inventory Management Journal*, 35 (1994), pp. 72-75. [K]

Billesbach, T., Harrison, A., and Croom-Morgan, S. "Just-In-Time: A United States-United Kingdom Comparison," *International Journal of Operations and Production Management*, 11 (1991), pp. 44-57. [C]

Billesbach, T., Harrison, A., and Morgan, S. C. "Just In Time: A United States-United Kingdom Comparison," *International Journal of Operations and Production Management*, 11 (1991), pp. 44-57. [I, E]

Billesbach, T., Harrison, A., and Morgan, S. C. "Supplier Performance Measures and Practices in JIT Companies in the U.S. and the U.K.," *International Journal of Purchasing and Materials Management*, 27 (1991), pp. 24-28. [P, E, I]

Billesbach, T., and Hayen, R. "Long-Term Impact of Just-In-Time on Inventory Performance Measures," *Production and Inventory Management*, 35 (1994), pp. 62-67. [J, N]

Billesbach, T., and Schniederjans, M. J. "Applicability of Just-In-Time Techniques in Administration," *Production and Inventory Management Journal*, 30 (1989), pp. 40-45. [C]

Bitran, G. R., and Chang, L. "A Mathematical Programming Approach to a Deterministic Kanban System," *Management Science*, 33 (1987), pp. 427-441. [K, N]

Bookbinder, J. H., and Locke, T. D. "Simulation Analysis of Just In Time Distribution," *International Journal of Physical Distribution and Materials Management*, 16 (1986), pp. 31-46. [S, P]

Bowden, R. O., Hall, J. D., and Usher, J. M. "Integration of Evolutionary Programming and Simulation to Optimize a Pull Production System," *Computers and Industrial Engineering*, 31 (1996), pp. 217-220. [S, N]

Boyer, K. K. "An Assessment of Managerial Commitment to Lean Production," *International Journal of Operations and Production Management*, 16 (1996), pp. 48-59. [C, H]

Brown, K. L. and Inman, R. Anthony "Small Business and JIT: A Managerial Overview," *International Journal of Operations and Production Management*, 13 (1993) pp. 57-66. [H, C]

Brown, K. A., and Mitchell, T.R. "A Comparison of Just-In-Time and Batch Manufacturing: The Role of Performance Obstacles," *Academy of Management Journal*, 34 (1991), pp. 906-917. [K]

Brox, J. A., and Fader, C. "Assessing the Impact of JIT Using Economic Theory," *Journal of Operations Management*, 15 (1998), pp. 371-378. [J]

Burbidge, J. L., Falster, P., and Riis, J. O. "Why is it Difficult to Sell GT and JIT to Industry?," *Production Planning and Control*, 2 (1991), pp. 160-166. [C]

Burnham, O. M. "Some Conclusions About JIT Manufacturing," *Production and Inventory Management Journal*, 28 (1987), pp. 7-10. [C]

Burton, T.T. "JIT/Repetitive Sourcing Strategies: Tying the Knot with your Suppliers," *Production and Inventory Management Journal*, 29 (1988), pp. 38-41. [H]

Butler, Dan "Here Today, Wrong Tomorrow," *Accountancy*, 121 (1998), pp. 40-42. [C]

Buxey, G., and Petzall, S. "Australian Automobile Industrial: JIT Production and Labor Relations," *Industrial Management and Data Systems*, 91 (1991), pp. 8-17. [I, H]

Cadley, J. A., Heintz, H. E., and Allocco, L.V. "Insights from Simulating JIT Manufacturing," *Interfaces*, 19 (1989), pp. 88-97. [S]

Callingham, M., and Smith, G. "Quality Comes to the Market Research World. Just In Time, or Just Too Late?," *Journal of the Market Research Society*, 36 (1994), pp. 269-293. [Q]

Campbell, J. F. "Materials Receiving Capacity and Inventory Management," *Omega*, 19 (1991), pp. 559-566. [C]

Caputo, M., and Dulmin, R. "Just-In-Time in ATO Vehicle Manufacturing: The Minivan-Piaggio V. E. Case Study," *Production Planning and Control*, 8 (1997), pp. 285-296. [C]

Caramanis, M., and Liberopoulos, G. "Perturbation Analysis for the Design of Flexible Manufacturing System Flow Controllers," *Operations Research*, 40 (1992), pp. 1107-1125. [N]

Carlson, J. G., and Yao, A. C. "Mixed Model Assembly Simulation," *International Journal of Production Economics*, 26 (1992), pp. 161-167. [S]

Carson, J.G. "JIT Applications to Warehousing Operations," *Engineering Costs and Production Economics*, 17 (1989), pp. 315-322. [P]

Celley, A. F., Clegg, W. H., Smith, A. W., and Vonderembse, M. A. "Implementation of JIT in the United States," *International Journal of Purchasing and Materials Management*, 22 (1986), pp. 9-15. [C]

Chakravorty, S. S., and Atwater, J. B. "A Comparative Study of Line Design Approaches for Serial Production Systems," *International Journal of Operations and Production Management*, 16 (1996), pp. 91-108. [C]

Chakravorty, S. S., and Atwater, J. B. "Do JIT Lines Perform Better than Traditionally Balanced Lines?" *International Journal of Operations and Production Management*, 15 (1995), pp. 77-88. [C]

Chan, F. T., and Smith, A. M. "Simulation Aids JIT Assembly Line Manufacture: A Case Study," *International Journal of Operations and Production Management*, 13 (1993), pp. 50-73. [S]

Chandrashekar, A. "Just-In-Time: A Case Study of the Grind Behind the Glamour," *International Journal of Purchasing and Materials Management*, 30 (1994), pp. 20-26. [C]

Chang, D., and Lee, S. M. "The Impact of Critical Success Factors of JIT Implementation on Organizational Performance," *Production Planning and Control*, 7 (1996), pp. 329-338. [J, C]

Chang, D. and Lee, S. M. "Impact of JIT on Organizational Performance of U. S. Firms," *International Journal of Production Research*, 33 (1995), pp. 3053-3068. [E, H]

Chang, F. C. "Heuristics for Dynamic Job Shop Scheduling with Real-Time Updated Queuing Time Estimates," *International Journal of Production Research*, 35 (1997), pp. 651-655. [N, K]

Changchit, C., and Kung, H. "Effect of Learning JIT Production System: A Simulation Experiment on MicroComputer," *Computers and Industrial Engineering*, 15 (1988), pp. 172-178. [E]

Chapman, S. N. "Just In Time Supplier Inventory: An Empirical Implementation Model," *International Journal of Production Research*, 27 (1989), pp. 1993-2007. [P]

Chapman, S. N. "Schedule Stability and the Implementation of Just-In-Time," *Production and Inventory Management Journal*, 31 (1990), pp. 66-70. [K]

Chapman, S. N. "Using Risk-Averse Inventory for JIT Process Improvements," *Production and Inventory Management Journal*, 33 (1992), pp. 69-74. [C, N]

Chapman, S. N., and Carter, P. L. "Supplier/Customer Inventory Relationships Under Just In Time," *Decision Sciences*, 21 (1990), pp. 35-51. [P]

Chaturvedi, M., and Golhar, D.Y. "Simulation Modeling and Analysis of a JIT Production System," *Production Planning and Control*, 3 (1992), pp. 81-92. [S]

Chaudhurg, A., and Whinston, B. "Toward an Adoptive Kanban System," *International Journal of Production Research*," 20 (1990), pp. 437-456. [K]

Chen, H. G. "Operator Scheduling Approaches in Group Technology Cells Information Request Analysis," *IEEE Transactions on Systems, Man, and Cybernetics*, 25 (1995), pp. 438-452. [K]

Chen, I. J., Chung, C. S., and Gupta, A. "The Integration of JIT and FMS: Issues and Decisions," *Integrated Manufacturing Systems*, 5 (1994), pp. 4-13. [C]

Chen, S., and Chen, R. "Manufacturer-Supplier Relationship in a JIT Environment," *Production and Inventory Management Journal*, 38 (1997), pp. 58-64. [P]

Cheng, D. W. "Line Reversibility of Tandem Queues with General Blocking," *Management Science*, 41 (1995), pp. 864-873. [N]

Cheng, L., and Ding, F. Y. "Modifying Mixed-Model Assembly Line Sequencing Methods to Consider Weighted Variations for Just-In-Time Production Systems," *IIE Transactions*, 28 (1996), pp. 919-927. [N]

Cheng, T. "Just In Time Production: A Survey of It's Development and Perception in Hong Kong Electronics Industry," *Omega*, 16 (1988), pp. 25-32. [I]

Cheng, T. "A Note on an Economic Production Quantity Model and Just-In-Time Production," *Production Planning and Control*, 4 (1993), pp. 88-92. [J]

Cheng, T. "Some Thoughts on the Practice of Just-In-Time Manufacturing," *Production Planning and Control*, 2 (1991), pp. 167-178. [C]

Cheng, T., and Musaphir, H. "Some Implementation Experiences with Just-In-Time Manufacturing," *Production Planning and Control*, 4 (1993), pp. 179-192. [C]

Chengalvarayn, G., and Parker, S. C. "Simulation of Just In Time Feasibility in Manufacturing Environment," *Computers and Industrial Engineering*, 21 (1991), pp. 303-306. [S]

Chin, L., and Rafuse, B. A. "A Small Manufacturer Adds JIT Techniques to MRP. (Just-In-Time; Manufacturing Resource Planning)," *Production and Inventory Management*, 34 (1993), pp. 18-21. [C, K]

Chu, C. H., and Shih, W. L. "Simulation Studies in JIT Production," *International Journal of Production Research*, 30 (1992), pp. 2573-2586. [S]

Chua, R. C. "Inventory Control of Purchased Materials When It's Not Quite JIT," *Production and Inventory Management Journal*, 33 (1992), pp. 14-18. [C]

Chyr, F., Lin, T. M., and Ho, C. F. "Comparison Between Just-In-Time and EOQ System," *Engineering Costs and Production Economics*, 18 (1990), pp. 233-250. [J, N]

Clarke, B., and Mia, L. "JIT Manufacturing Systems: Use and Application in Australia," *International Journal of Operations and Production Management*, 13 (1993), pp. 69-82. [I]

Clinton, D., and Hsu, K. C. "JIT and the Balanced Scorecard: Linking Manufacturing Control to Management Control," *Management Accounting*, 79 (1997), pp. 18-21. [C]

Clouse, V. G. H., and Gupta, Y. P. "Just In Time and the Trucking Industry: Implications of the Motor Carrier Act," *Production and Inventory Management Journal*, 30 (1990), pp. 7-12. [P]

Co, H. C., and Jacobson, S. H. "The Kanban Assignment Problem in Serial Just-In-Time Production Systems," *IIE Transactions*, 26 (1994), pp. 76-85. [K]

Co, H. C., and Sharafali, M. "Over Planning Factor in Toyota's Formula for Computing the Number of Kanbans," *IIE Transactions*, 29 (1997), pp. 409-415. [K]

Co, H. C., and Zhu, J. "A Note on the Problem of Scheduling a Flexible Manufacturing System for Just-In-Time Customers," *International Journal of Production Research*, 33 (1995), pp. 1673-1681. [K]

Coleman, B. J., and Jennings, Kenneth M. "The UPS Strike: Lessons for Just-In-Timers," *Production & Inventory Management Journal*, 39 (1998), pp. 63-68. [C, H]

Coleman, B. J., Vaghefi, M., and Reza, H. "A Key to the Toyota Production System," *Production and Inventory Management*, 35 (1994), pp. 31-35. [C]

Collins, R., Bechler, K., and Pires, S. "Outsourcing in the Automotive Industry: From JIT to Modular Consortia," *European Management Journal*, 15 (1997), pp. 498-502. [P]

Cook, David P. "A Simulation Comparison of Traditional, JIT, and TOC Manufacturing Systems in a Flow Shop with Bottlenecks. (Just-In-Time; Theory of Constraints)," *Production and Inventory Management Journal*, 35 (1994), pp. 73-78. [S]

Cook, R. L., and Rogowski, R. A. "Applying JIT Principles to Continuous Process Manufacturing Supply Chains," *Production and Inventory Management Journal*, 37 (1996), pp. 12-16. [P]

Conti, R. F. "Variable Manning JIT: An Innovative Answer to Team," *Production and Inventory Management Journal*, 37 (1996), pp. 24-27. [H]

Corbey, M., and Jansen, R. "The Economic Lot Size and Relevant Costs," *International Journal of Production Economics*, 30 (1993), pp. 519-530. [N, J]

Cordon, C. "Quality Defaults and Work-In-Process Inventory," *European Journal of Operational Research*, 80 (1995), pp. 240-251. [Q]

Cormier, G., and Kersey, D. F. "Conceptual Design of a Warehouse for Just-In-Time Operations in a bakery," *Computers and Industrial Engineering*, 29 (1995), pp. 361-365. [P]

Courtis, J. K. "JIT's Impact on a Firm's Financial Statements," *International Journal of Purchasing and Materials Management*, 31 (1995), pp. 46-50. [P]

Cowton, C. J., and Vail, R. L. "Making Sense of Just-In-Time Production: A Resource-Based Perspective," *Omega*, 22 (1994), pp. 427-441. [C]

Crawford, K. M., Blacktone, J. H., and Cox, J. F. "A Study of JIT Implementation and Operating Problems," *International Journal of Production Research*, 26 (1988), pp. 1561-1568. [C]

Crawford, K. M., and Cox, J. F. "Addressing Manufacturing Problems Through the Implementation of Just In Time," *Production and Inventory Management Journal*, 31 (1991), pp. 33-38. [C]

Crawford, K. M., and Cox, J. F. "Designing Performance Measurement Systems for Just-In-Time Operations," *International Journal of Production Research*, 28 (1990), pp. 2025-2036. [N, J]

Crandall, R. E., and Burwell, T. "The Effect of Work-In-Process Inventory Levels on Throughput and Lead Times," *Production and Inventory Management Journal*, 34 (1993), pp. 6-12. [C]

Crosby, L. "The Just In Time Manufacturing Process: Control of Quality and Quantity," *Production and Inventory Management Journal*, 25 (1984), pp. 21-33. [Q]

Cusumano, M. A. "The Limits of Lean," *Sloan Management Review*, 35 (1994), pp. 27-32. [C]

Dahlén, P., Ericsson, J., and Fujii, H. "Labor Stability and Flexibility -- Conditions to Reach Just-In-Time," *International Journal of Operations and Production Management*, 15 (1995), pp. 26-43. [H]

Daniel, S. J., and Reitsperger, W. D. "Management Control Systems for JIT: An Empirical Comparison of Japan and the US," *Journal of International Business Studies*, 22 (1991), pp. 603-617. [I, E]

Das, S. K., and Bhambri, S. "A Decision Tree Approach for Selecting Between Demand Based, Reorder, and JIT/Kanban Methods for Material Procurement," *Production Planning and Control*, 5 (1994), pp. 342-348. [N, K, P]

Das, C., and Goyal, S. K. "A Vendor's View of the JIT Manufacturing System," *International Journal of Operations and Production Management*, 9 (1989), pp. 106-111. [P]

Daugherty, P. J., Rogers, D., and Spencer, M. S. "Just-In-Time Functional Model: Empirical Test and Validation," *International Journal of Physical Distribution and Logistics Management*, 24 (1994), pp. 20-26. [C, E]

Davy, J. A., White, R. E., Merritt, N. J., and Gritzmacher, K. "A Derivation of the Underlying Constructs of Just-In-Time Management Systems. (Research Notes)," *Academy of Management Journal*, 35 (1992), pp. 653-670. [E]

Dean Jr., James W., and Snell, S. A. "The Strategic Use of Integrated Manufacturing: An Empirical Examination," *Strategic Management Journal*, 17 (1996), pp. 459-480. [C]

Delbridge, R. "Surviving JIT: Control and Resistance in a Japanese Transplant," *Journal of Management Studies*, 32 (1995), pp. 803-817. [C]

Delbridge, R., and Olive, N. "Narrowing the Gap? Stock Turns in the Japanese and Western Car Industry," *International Journal of Production Research*, 29 (1991), pp. 2083-2095. [C]

Deleersnyder, J. L., Hodgson, T. J., King, R. E., O'Grady, P. J., and Savva, A. "Integrating Kanban Type Pull Systems and MRP Type Push Systems: Insights From a Markovian Model," *IIE Transactions*, 24 (1992), pp. 43-56. [K]

Deleersnyder, J. L., Hodgson, T. J., Muller, H., and O'Grady, P. J. "Kanban Controlled Pull Systems: An Analytic Approach," *Management Science*, 35 (1989), pp. 1079-1091. [K]

Deshpande, S. P., and Golhar, D. Y. "HRM Practices in Unionized and Nonunionized Canadian JIT Manufacturing Firms," *Production and Inventory Management Journal*, 36 (1995), pp. 15-19. [H, E, I]

Deshpande, S. P., and Golhar, D.Y. "HRM Practices of JIT firms in Canada," *Production Planning and Control*, 7 (1996), pp. 79-85. [H, I]

Deshpande, S. P., Golhar, D.Y., and Stamm, C. L. "Human Resource Management in the Just-In-Time Environment," *Production Planning and Control*, 5 (1994), pp. 372-380. [H]

Ding, F. Y. "Kitting in Just In Time Production," *Production and Inventory Management Journal*, 30 (1990), pp. 25-32. [C]

Ding, F. Y., and Cheng, L. "An Effective Mixed-Model System Assembly Line Sequencing Heuristic for Just-In-Time Productions Systems," *Journal of Operations Management*, 11 (1993), pp. 45-50. [N, K]

Ding, F. Y., and Yuen, M. N. "A Modified MRP for a Production System with the Coexistence of MRP and Kanbans," *Journal of Operations Management*, 10 (1991), pp. 267-277. [K]

Dion, P. A., Banting, P. M., Picard, S., and Blenkhorn, D. L. "JIT Implementation: A Growth Opportunity for Purchasing," *International Journal of Purchasing and Materials Management*, 28 (1992), pp. 32-38. [P]

Discenza, R., and McFadden, F. R. "The Integration of MRP II and JIT Through Software Unification," *Production and Inventory Management Journal*, 29 (1988), pp. 49-53. [K]

Dixon, Lance "Got a Problem? Get JIT II," *Purchasing*, 123 (1997), pp. 31-32. [P]

Dixon, Lance "JIT II: The Ultimate Customer-Supplier Partnership," *Hospital Management Quarterly*, 20 (1999), pp. 14-21. [P]

Dong, H. J., Hayya, J. C., and Lae, K. S. "JIT Purchasing and Setup Reduction in an Integrated Inventory Model," *International Journal of Production Research*, 30 (1992), pp. 255-261. [P]

Dove, Rick "Creating & Communicating Agility Insights," *Automotive Manufacturing & Production*, 109 (1997), pp. 18-20. [C]

Dowlatshahi, S. "Implementing Early Supplier Involvement: A Conceptual Framework," *International Journal of Operations & Production Management*, 18 (1998), pp. 143-168. [C]

D'Ouville, E., Willis, T., and Huston, C.R. "A Note on the EOQ-JIT Relationship," *Production Planning and Control*, 3 (1992), pp. 57-60. [J, N]

Duenyas, I., Hopp, W., and Bassok, Y. "Production Quotas as Bounds on Interplant Contracts," *Management Science*, 43 (1997), pp. 1372-1378. [C]

Duenyas, I., and Keblis, M. F. "Release Policies for Assembly Systems," *IIE Transactions*, 27 (1995), pp. 507-518. [C]

Ebrahimpour, M., and Fathi, B. M. "Dynamic Simulation of Kanban Production Inventory System," *International Journal of Operations and Production Management*, 5 (1984), pp. 5-14. [S, K]

Ebrahimpour, M., and Schoenberger, R. J. "The Japanese Just In Time/Total Quality Control Production System," *International Journal of Production Research*, 22(1984), pp. 422-427. [Q]

Ebrahimpour, M., and Withers, B. E. "A Comparison of Manufacturing Management in JIT and Non-JIT Firms," *International Journal of Production Economics*, 32 (1993), pp. 355-364. [E]

Edwards, D. K., Edgell, R. C., and Richa, C. E. "Standard Operations - The Key to Continuous Improvement in a Just-In-Time Manufacturing System," *Production and Inventory Management Journal*, 34 (1993), pp. 7-13. [C]

Ehrhardt, R. "A Model of JIT Make-To-Stock Inventory with Stochastic Demand," *The Journal of the Operational Research Society*, 48 (1997), pp. 1013-1017. [N]

Erdem, A., and Swifit, Cathy Owens "Items to Consider for Just-In-Time Use in Marketing Channels: Toward a Development of a Decision Tool," *Industrial Marketing Management*, 27 (1998), pp. 21-30. [P]

Ericsson, Johan, and Dahlen, Per "A Conceptual Model for Disruption Causes: A Personnel Organization Perspective," *International Journal of Production Economics*, 52 (1997), pp. 47-54. [H]

Ericsson, J., and Fujii, H. "Labor Stability and Flexibility - Conditions to Reach Just-In-Time," *International Journal of Operations and Production Management*, 15 (1995), pp. 26-43. [H]

Esparrago, J. R. "Kanban," *Production and Inventory Management Journal*, 29 (1988), pp. 6-10. [K]

Estrada, F., Villalobos, J. R., and Roderick, L. "Evaluation of Just-In-Time Alternatives in the Electric Wire-Harness Industry," *International Journal of Production Research*, 35 (1997), pp. 1993-2008. [C]

Etkin, J. "Real-Time Network in the Manufacturing Plant: Intelligence, Not Just a Link," *Robotics and Computer-Integrated Manufacturing*, 6 (1989), pp. 191-198. [C]

Ezingeard, J. N., and Race, P. "Spreadsheet Simulation to Aid Capacity Management of Batch Chemical Processing Using JIT Pull Control," *International Journal of Operations and Production Management*, 15 (1995), pp. 82-88. [S]

Fallon, D., and Browne, J. "Simulating Just In Time Systems," *International Journal of Operations and Production Management*, 8 (1988), pp. 30-45. [S]

Fandel, G., and Reese, J. "Just-In-Time Logistics of a Supplier in the Car Manufacturing Industry," *International Journal of Production Economics*, 24 (1991), pp. 55-64. [P]

Fawcett, S. E., and Birou, L. M. "Exploring the Logistics Interface Between Global and JIT Sourcing," *International Journal of Physical Distribution and Logistics Management*, 22 (1992), pp. 3-15. [P]

Fawcett, S. E., and Birou, L. M. "Just-In-Time Sourcing Techniques: Current State of Adoption and Performance Benefits," *Production and Inventory Management Journal*, 34 (1993), pp. 18-24. [P]

Fawcett, S. E., and Scully, J. "A Contingency Perspective of Just-In-Time Purchasing: Globalization, Implementation and Performance," *International Journal of Production Research*, 33 (1995), pp. 915-931. [P]

Fazel, F. "A Comparative Analysis of Inventory Costs and EOQ Purchasing," *International Journal of Physical Distribution and Logistics*, 27 (1997), pp. 496-505. [J, N]

Fazel, F., Fischer, K. and Gilbert, E. W. "JIT Purchasing vs. EOQ with a Price Discount: An Analytical Comparison of Inventory Costs," *International Journal of Production Economics*, 54 (1998), pp. 101-109. [J, N, P]

Federgruen, A., and Mosheiov ,G. "Heuristics for Multimachine Scheduling Problems with Earliness and Tardiness Costs," *Management Science*, 42 (1996), pp. 1544-1533. [K]

Feller, G. "Use of GPSS/PC to Establish Manning Level of Proposed Just In Time Production Facility," *Computers and Industrial Engineering*, 11 (1984), pp. 382-384. [S]

Ferguson, P. "General Managers in the JIT Cross Fire," *Journal of Management*, 14 (1989), pp. 5-17. [H, C]

Fiedler, K., Galletly, J. E., and Bicheno, J. "Expert Advice for JIT Implementation," *International Journal of Operations and Production Management*, 13 (1993), pp. 23-30. [C]

Fieten, R. "Integrating Key Suppler - Essential Part of Just In Time Concept," *Engineering Cost and Production Economics*, 15 (1989), pp. 185-189. [P]

Finch, B. "Japanese Management Techniques in Small Manufacturing Companies: A Strategy for Implementation," *Production and Inventory Management Journal*, 27 (1986), pp. 30-37. [C]

Finch, B. "JIT, TOC, and BPR: An Overview of Productivity Improvement Resources on the Internet," *Production and Inventory Management Journal*, 37 (1996), pp. 86-88. [Q, N]

Finch, B., and Cox, F. "An Examination of Just In Time Management for the Small Manufacturer: With an Illustration," *International Journal of Production Research*, 24 (1991), pp. 329-342. [C]

Flapper, S. D., Miltenberg, G., and Wijngaard, J. "Embedding JIT into MRP," *International Journal of Production Research*, 29 (1991), pp. 329-341. [K]

Flynn, B. B., Sakakibara, S., and Schroeder, R. G. "Relationship Between JIT and TQM Practices and Performance," *Academy of Management Journal*, 38 (1995), pp. 1325-1360. [Q]

Fojt, M. "Measuring the Success of Just In Time Production," *International Journal of Physical Distribution and Logistics Management*, 25 (1995), p. 2. [C]

Fraizer, G. L., Spekman, R. E., and O'Neal, C.R. "Just In Time Exchange Relationships in Industrial Markets," *Journal of Marketing*, 52 (1988), pp. 52-67. [C]

Freeland, J. R., "A Survey of Just-In-Time Purchasing Practices in the United States," *Production and Inventory Management*, 32 (1991), pp. 43-49. [C]

Freeland, J. R., Leschke, J., and Weiss, E. "Guidelines for Setup-Cost Reduction Programs to Achieve Zero Inventory," *Journal of Operations Management*, 9 (1990), pp. 85-100. [C]

Frein, Y., Mascolo, M. D., and Dallery, Y. "On the Design of Generalized Kanban Control Systems," *International Journal of Operations and Production Management*, 15 (1995), pp. 158-184. [K]

Freudmann, Aviva "British Consultants Tell US Firms to Get a Grip – On Supply Chain," *Journal of Commerce and Commercial*, 415 (1998), pp. 1-3. [P]

Fujimura, K. "Time-Minimum Routes in Time-Dependent Networks," *IEEE Transactions on Robotics and Automation*, 11 (1995), pp. 343-351. [P]

Fukukawa, T., and Hong, S. C. "The Determination of the Optimal Number of Kanbans in a Just-In-Time Production System," *Computers and Industrial Engineering,* 24(1993), pp. 551-560. [K, N]

Funk, J. L. "Just-In-Time Manufacturing and Logistical Complexity: A Contingency Model," *International Journal of Operations and Production Management*, 15 (1995), pp. 60-71. [P]

Gabriel, T., Bicheno, J., and Galletly, J. "JIT Manufacturing Simulation," *Industrial Management and Data Systems*, 91 (1991), pp. 3-7. [S]

Gagne, M. L., and Discenza, R. "Accurate Product Costing in a JIT Environment," *International Journal of Purchasing and Materials Management*, 28 (1992), pp. 28-31. [J, N]

Garg, D., Kaul, O. N., and Deshmukh, S. G. "JIT Implementation: A Case Study," *Production and Inventory Management Journal*, 39 (1998), pp. 26-33. [C]

Gerchak, Y., and Parlar, M. "Investing in Reducing Lead-Time Randomness in Continuous-Review Inventory Models," *Engineering Costs and Production Economics*, 21 (1991), pp. 191-197. [N]

Germain, R., and Droge, C. "The Context, Organizational Design, and Performance of JIT Buying Versus Non-JIT Buying Firms," *International Journal of Purchasing and Materials Management*, 32 (1998), pp. 127-134. [P]

Germain, R., and Droge, C. "Just-In-Time and Context: Predictors of Electronic Data Interchange Technology Adoption," *International Journal of Physical Distribution and Logistics Management*, 25 (1995), pp. 18-33. [C]

Germain, R., Droge, C., and Daugherty, P. "The Effect of Just-In-Time Selling on Organizational Structure: An Empirical Investigation," *Journal of Marketing Research,* 31 (1994), pp. 471-483. [E]

Ghalayini, A., Noble, J., and Crowe, T. "An Integrated Dynamic Performance Measurement System for Improving Manufacturing Competitiveness," *International Journal of Production Economics*, 48 (1997), pp. 207-225. [C]

Gilbert, J. "The State of JIT Implementation and Development in the USA," *International Journal of Production Research*, 28 (1990), pp. 1099-1110. [C]

Giust, L. "Just-In-Time Manufacturing and Material-Handling Trends," *International Journal of Physical Distribution and Logistics Management*, 23 (1993), pp. 32-38. [P]

Giust, L. "Just-In-Time Manufacturing and Material Handling Trends," *Industrial Management and Data Systems*, 93 (1993), pp. 3-9. [P]

Glasserman, Paul and Wang, Yashan "Leadtime-Inventory Trade-offs in Assemble-to-order Systems," *Operations Research*, 46 (1998), pp. 858-867. [K]

Goh, M., and Hum, S. "Cost Bounds for Inventory Systems Approaching JIT," *International Journal of Operations and Production Management*, 11 (1991), pp. 59-63. [J, N]

Goldhar, D. Y., and Stamm, C. L. "The Just In Time Philosophy: A Literature Review," *International Journal of Production Research*, 29 (1991), pp. 657-676. [C]

Golhar, D. Y., and Deshpande, S. P. "An Empirical Investigation of HRM Practices in JIT Firms," *Production and Inventory Management Journal*, 34 (1993), pp. 28-32. [H, E]

Golhar, D. Y., and Stamm, C. L. "JIT Purchasing Practices in Manufacturing Firms," *Production and Inventory Management Journal*, 34 (1993), pp. 75-79. [P]

Golhar, D.Y., Stamm, C. L., and Banerjee, S. "Just-In-Time Purchasing: An Empirical Investigation," *Production Planning and Control*, 4 (1993), pp. 392-398. [P]

Golhar, D. Y., Stamm, C., and Smith, W. "JIT Implementation in Small Manufacturing Firms," *Production and Inventory Management Journal*, 31 (1990), pp. 44-53. [C]

Golhar, D. Y., and Surrey, B. R. "Economic Manufacturing Quantity in a Just In Time Delivery System," *International Journal of Production Research*, 30 (1992), pp. 961-973. [J, N]

Gomes, R., and Mentzer, J. T. "A Systems Approach to the Investigation of Just In Time," *Journal of Business Logistics*, 9 (1988), pp. 71-88. [C]

Goyal, S. K. "Joint Vendor-Buyer Policy in JIT Manufacturing," *The Journal of the Operational Research Society*, 48 (1997), pp. 550-560. [P]

Goyal, S. K., and Deshmukh, S. G. "A Critique of the Literature on Just-In-Time Manufacturing," *International Journal of Operations and Production Management*, 12 (1992), pp. 18-28. [C]

Goyal, S. K., and Deshmukh, S. G. "Integrated Procurement-Production System in a Just-In-Time Environment: Modeling and Analysis," *Production Planning and Control*, 8 (1997), pp. 31-36. [C]

Grande, P. D., and Satir, A. "Adoption of Just-In-Time Based Quality Assurance and Purchasing Practices," *Production Planning and Control*, 5 (1994), pp. 397-406. [Q, P]

Grant, M. R "EOQ and Price Break Analysis in a JIT Environment," *Production and Inventory Management Journal*, 34 (1993), pp. 64-68. [J, N]

Gravel, M., and Price, W. L. "Visual Interactive Simulation Shows How to Use the Kanban Method in Small Business," *Interfaces*, 21 (1991), pp. 22-31. [S, K, C]

Groebner, D. F., and Merz, C. "The Impact of Implementing JIT on Employees' Job Attitudes," *International Journal of Operations and Production Management*, 14 (1994), pp. 26-37. [H]

Grout, J. R. "A Model of Incentive Contracts for Just-In-Time Delivery," *European Journal of Operational Research*, 96 (1997), pp. 139-146. [P]

Grout, J. R. "Influencing a Supplier Using Delivery Windows: On the Variance of Flow Time and On-time Delivery," *Decision Sciences*, 29 (1998), pp. 747-456. [P, N]

Grout, J. R., and Seastrand, M. "Multiple Operation Lot Sizing in a Just In Time Environment," *Production and Inventory Management*, 28 (1987), pp. 23-26. [N]

Grunwald, H. J., and Fortuin, L. "Many Steps Toward Zero Inventory," *European Journal of Operational Research*, 59 (1992), pp. 359-369. [C]

Giunipero, L. C. "Motivating and Monitoring JIT Supplier Performance (Just In Time Systems)," *Journal of Purchasing and Materials Management*, 26 (1990), pp. 19-25. [P]

Giunipero, L. C., and Law, W. K. "Organizational Changes and JIT Implementation," *Production and Inventory Management*, 31 (1990), pp. 71-73. [H, C]

Gunasekaran, A., Goyal, S. K., Martikainen, T., and Yli-Olli, P. "Determining Economic Inventory Policies in a Multi-stage Just-In-Time Production System," *International Journal of Production Economics*, 30 (1993), pp. 531-542. [J, N]

Gunasekaran, A., Goyal, S. K., Martikainen, T., and Yli-Olli, P. "Equipment Selection in Just-In-Time Manufacturing Systems," *Journal of the Operational Research Society*, 44 (1993), pp. 345-353. [C]

Gunasekaran, A., Goyal, S. K., Martikainen, T., and Yli-Olli, P. "Modeling and Analysis of Just-In-Time Manufacturing Systems," *International Journal of Production Economics*, 32 (1993), pp. 23-38. [N]

Gunasekaran, A., and Lyu, J. "Implementation of Just-In-Time in a Small Company: A Case Study," *Production Planning and Control*, 8 (1997), pp. 406-412. [C]

Gunter, F., and Reese, J. "Just In Time Logistics of Suppliers in the Car Manufacturing Industry," *International Journal of Operation Production Economics*, 24 (1991), pp. 55-64. [P]

Gupta, S., and Brennan, L. "A Knowledge Based System for Combined Just-In-Time and Material Requirements Planning," *Computers and Electrical Engineering*, 19 (1993), pp. 157-174. [K]

Gupta, O., and Kini, R. B. "Is Price-Quantity Discount Dead in a Just-In-Time Environment? *International Journal of Operations and Production Management*, 15 (1995), pp. 261-270. [J, N]

Gupta, S., and Brennan, L. "Implementation of Just-In-Time Methodology in a Small Company," *Production Planning and Control*, 6 (1995), pp. 358-364. [C]

Gupta, Y. "A Feasibility Study of JIT Purchasing Implementation in a Manufacturing Facility," *International Journal of Operations and Production Management*, 20 (1990), pp. 31-41. [P]

Gupta, Y. P., and Bagchi, P. K. "Inbound Freight Consolidation Under Just In Time Procurement: Application of Clearing Models," *Journal of Business Logistics*, 8 (1987), pp. 74-94. [P]

Gupta, Y. P., and Gupta, M. "A System Dynamics Model of a JIT-Kanban System," *Engineering Costs and Production Economics*, 18 (1989), pp. 117-130. [N, K]

Gupta, Y. P., Lonial, S. C., and Mangold, W. G. "Empirical Assessment of the Strategic Orientation of JIT Manufacturers Versus Non-JIT Manufacturers," *Computer-Integrated Manufacturing Systems*, 5 (1992), pp. 181-190. [E]

Gutiérrez, R. A., and Sahinidis, N. V. "A Branch-and-Bound Approach for Machine Selection in Just-In-Time Manufacturing Systems," *International Journal of Production Research*, 34 (1996), pp. 797-818. [N]

Haan, J. de, and Yamamoto, Masaru M. "Zero Inventory Management: Facts or Fiction? Lessons from Japan," *International Journal of Production Economics*, 59 (1999), pp. 65-66. [J, P]

Ha, D., and Kim, S. L. "Implementation of JIT Purchasing: An Integrated Approach," *Production Planning and Control*, 8(1997), pp. 152-158. [P]

Hahm, J., and Yano, C.A. "The Economic Lot and Delivery Scheduling Problem: The Common Cycle Case," *IIE Transactions*, 27 (1995), pp. 113-125. [C]

Hahm, J., and Yano, C. A. "The Economic Lot and Delivery Scheduling Problem: Models for Nested Schedules," *IIE Transactions*, 27 (1995), pp. 126-139. [N, K]

Hahn, C., Pinto, P.A., and Bragg, D. J. "Just In Time Production and Purchasing," *Journal of Purchasing and Materials Management*, 19 (1983), pp. 2-10. [P]

Halim, A., Miyazaki, S., and Ohta, H. "Batch-Scheduling Problems to Minimize Actual Flow Times of Parts through the Shop Under JIT Environment. (Just-In-Time)," *European Journal of Operational Research*, 72 (1994), pp. 529-544. [K]

Halim, A., Miyazaki, S., and Ohta, H. "Lot Scheduling Problems of Multiple Items in the Shop with both Receiving and Delivery Just In Time," *Production Planning and Control*, 5 (1994), pp. 175-184. [K]

Halim, A., and Ohta, H. "Batch Scheduling Problems through the Flowshop with both Receiving and Delivery Just-In-Time," *International Journal of Production Research*, 31 (1993), pp. 1943-1955. [K]

Halim, A., and Ohta, H. "Batch-Scheduling Problems to Minimize Inventory Cost in the Shop with both Receiving and Delivery Just-In-Time," *International Journal of Production Economics*, 33 (1994), pp. 185-194.

Hall, John D., Bowen, Roy, Grant, Richard S., and Hadley, William H. "An Optimizer for the Kanban Sizing Problem: A Spread Application for Whirlpool Corporation," *Production & Inventory Management Journal*, 39 (1998), pp. 17-23. [K, N]

Hallihan, A., Sackett, P., and Williams, G. M. "JIT Manufacturing: the Evolution to an Implementation Model Founded in Current Practice," *International Journal of Production Research*, 35 (1997), pp. 901-921. [C]

Handfield, R. "Distinguishing Features of Just-In-Time Systems in the Make-To-Order/Assemble-to-Order Environment. *Decision Sciences*, 24 (1993), pp. 581-601. [N, C]

Hannah, K., "Just In Time: Meeting the Competitive Challenge," *Production and Inventory Management*, 28 (1987), pp. 1-3. [C]

Harber, D., Samson, D. A., Sohal, A., and Wirth, A. "Just In Time: The Issue of Implementation," *International Journal of Production Management*, 10 (1990), pp. 21-30. [C]

Harris, E. "The Impact of JIT Production on Product Costing Information Systems," *Production and Inventory Management Journal*, 31 (1990), pp. 44-51. [C]

Harrison, A., and Voss, C. "Issues in Setting up JIT Supply. *International Journal of Operations and Production Management*, 10 (1990), pp. 84-93. [P, C]

Haynsworth, H. C. "A Theoretical Justification for the Use of Just-In-Time Scheduling," *Production and Inventory Management*, 25 (1984), pp. 1-3. [J]

Hazra, J., Schweitzer, P. J., and Seidmann, A. "Analyzing Closed Kanban-Controlled Assembly Systems by Interative Aggregation-Disaggregation," *Computers & Operations Research*, 26 (1999), pp. 1015-1039. [K, N]

Horn, G. Scott, and Cook, Robert Lorin "Heijunka Transportation Measure: Development and Application," *Production Inventory Management Journal*, 38 (1997), pp. 32-39. [C]

Hedin, S. R., and Russell, G. R. "JIT Implementation: Interaction Between the Production and Cost Accounting Functions," *Production and Inventory Management*, 33 (1992), pp. 68-73. [J, C]

Hegstad, M. "A Simple, Low-Risk, Approach to JIT," Production and Planning Control, 1 (1990), pp. 53-64. [C]

Heiko, L. "Simple Framework for Understanding JIT," *Production and Inventory Management*, 30 (1990), pp. 61-63. [C]

Helms, M. M. "Communication: The Key to JIT Success," *Production and Inventory Management*, 31 (1990), pp. 18-21. [C]

Hendrick, T. E. "The Pre-JIT/TQC Audit: First Step of the Journey," *Production and Inventory Management*, 28 (1987), pp. 132-142. [C, Q]

Hill, A. V., and Vollman, T. E. "Reducing Vendor Delivery Uncertainties in a JIT Environment," *Journal of Operations Management*, 6 (1986), pp. 381-392. [P, N]

Hobbs Jr., O. "Application of JIT Techniques in a Discrete Batch Job Shop," *Production and Inventory Management Journal*, 35 (1994), pp. 43-47. [K]

Hobbs Jr., O. "Managing JIT Toward Maturity," *Production and Inventory Management Journal*, 38 (1997), pp. 47-50. [C]

Hoeffer, E. L. "GM tries Just-In-Time vs. Just In Case Production/Inventory Systems: Concepts Borrowed back from Japan," *Production and Inventory Management*, 23 (1982), pp. 12-20. [C]

Hoffman, C. "Comments of 'Joint Vendor-Buyer Policy in JIT Manufacturing,'" *The Journal of the Operational Research Society*, 48 (1997), pp. 546-555. [P]

Holdsworth, R., and Dale, B. G. "A Just-In-Time Production System: A Case Study," *International Journal of Manufacturing Systems Design*, 2 (1995), pp. 51-60. [C]

Hong, J. D., and Hayya, J. C. "Joint Investment in Quality Improvement and Setup Reduction," *Computers and Operations Research*, 22 (1995), pp. 567-574. [Q]

Hong, J. D., and Hayya, J. C. "Just-In-Time Purchasing: Single or Multiple Sourcing?" *International Journal of Production Economics*, 27 (1992), pp. 175-181. [P]

Hong, J. D., Hayya, J. C., and Kim, S. L. "JIT Purchasing and Setup Reduction in an Integrated Inventory Model," *International Journal of Production Research*, 30 (1992), pp. 255-266. [P, N]

Hoogeveen, J. A., Oosterhout, H., and Van De Velde, S. L. "New Lower and Upper Bounds for Scheduling Around a Small Common Due Date," *Operations Research*, 42 (1994), pp. 102-110. [K, N]

Houghton, E., and Portougal, V. "A Planning Model for Just-In-Time batch Manufacturing," *International Journal of Operations and Production Management*, 15 (1995), pp. 9-25. [K]

Howard, M. L., and Newman, R. G. "From Job Shop to Just-In-Time - A Successful Conversion," *Production and Inventory Management*, 34 (1993), pp. 70-74. [C]

Huang, P.Y., Rees, P. L., and Taylor, B.W. "A Simulation Analysis of Japanese JIT Multi-Machine, Multi-Stage Production System," *Decision Sciences*, 14 (1983), pp. 326-343. [S]

Hudson, R., and Sadler, D. "'Just-In-Time' Production and the European Automotive Components Industry," *International Journal of Physical Distribution and Logistics Management*, 22 (1992), pp. 40-45. [I]

Hum, S. H. "Industrial Progress and the Strategic Significance of JIT and TQC for Developing Countries. *International Journal of Operations and Production Management*, 11 (1991), pp. 39-46. [Q, C]

Hum, S. H., and Lee, C. K. "JIT Scheduling Rules: A Simulation Evaluation," *Omega*, 26 (1998), pp. 381-387. [K, S]

Hum, S. H., and Ng, Y. T. "A Study on Just-In-Time Practices in Singapore," *International Journal of Operations and Production Management*, 15 (1995), pp. 5-24. [I]

Hunglin, W., and Wang, H. P. "Optimum Number of Kanbans between Two Adjacent Workstations in JIT System," *International Journal of Production Economics*, 22 (1991), pp. 179-188. [K, N]

Huq, Z., and Huq, F. "Embedding JIT in MRP: The Case of Job Shops," *Journal of Manufacturing Systems*, 13 (1994), pp. 153-154. [K]

Huq, F., and Pinney, W. E. "Impact of Short-Term Variations in Demand on Opportunity Costs in a Just-In-Time Kanban System," *Production and Inventory Management Journal*, 32 (1991), pp. 8-13. [K]

Hurrion, R. D. "An Example of Simulation Optimisation Using a Neural Network Metamodel: Finding the Optimum Number of Kanbans in Manufacturing Systems," *Journal of the Operations Research Society*, 48 (1997), pp. 1105-1113. [S, N, K]

Iijima, M., Komatsu., S., and Katoh, S. "Hybrid Just-In-Time Logistics Systems and Information Networks for Effective Management in Perishable Food

Industries. *International Journal of Production Economics*, 44 (1996), pp. 97-103. [P]

Im, J. "How Does Kanban Work in American Companies," *Production and Inventory Management Journal*, 30 (1989), pp. 22-25. [K]

Im, J. "Lessons from Japanese Production Management," *Production and Inventory Management Journal*, 30 (1989), pp. 25-33. [C]

Im, J., Hartman, S., and Bondi, P. "How Does JIT Systems Affect Human Resource Management?" *Production and Inventory Management Journal*, 35 (1994), pp. 1-4. [H]

Im, J., and Lee, S. M. "Implementation of Just-In-Time Systems in US Manufacturing Firms. *International Journal of Operations and Production Management*, 9 (1989), pp. 5-14. [C]

Im, J., and Schoenberger, R. "The Pull of Kanban," *Production and Inventory Management Journal*, 29 (1988), pp. 54-57. [K]

Imai, Masaaki, and Allnoch, Allen "Q&A" *Industrial Management*, 40 (1998), pp. 4-7. [C]

Inman, R. A. "The Impact of Lot-Size Reduction on Quality," *Production and Inventory Management Journal*, 35 (1994), pp. 5-7. [Q]

Inman, R.A. "Quality Certification of Suppliers by JIT Manufacturing Firms," *Production and Inventory Management Journal*, 30 (1990), pp. 58-63. [Q]

Inman, R. A., and Boothe, R. "The Impact of Quality Circles on JIT Implementation," *International Journal of Quality and Reliability Management*, 10 (1993), pp. 7-15. [Q]

Inman, R. A., and Brandon, L.D. "An Undesirable Effect of JIT. (Just In Time, Job Stress)," *Production and Inventory Management Journal*, 33 (1992), pp. 55-58. [C]

Inman, R. A., and Mehra, S. "Financial Justification of JIT Implementation. *International Journal of Operations and Production Management*, 13 (1993), pp. 32-38. [J]

Inman, R. A., and Mehra, S. "Potential Union Conflict in JIT Implementation?" *Production and Inventory Management Journal*, 29 (1989), pp. 19-21. [H]

Inman, R. A., and Mehra, S. "The Transferability of Just-In-Time Concepts to American Small Businesses. *Interfaces*, 20 (1990), pp. 30-37. [C]

Inman, R. R., and Bulfin, R. L. "Quick and Dirty Sequencing for Mixed-Model Multi-Level JIT Systems," *International Journal of Production Research*, 30 (1992), pp. 2011-2018. [K]

Ittner, C. D. "Exploratory Evidence on the Behavior of Quality Costs," *Operations Research*, 44 (1996), pp. 114-130. [Q]

Jaber, M. Y., and Bonney, M. "The Economic Manufacture/Order Quantity (EMQ/EOQ) and the Learning Curve: Past, Present, and Future," *International Journal of Production Economics*, 59 (1999), pp. 93-94. [J]

Jamal, A., and Sarker, B. R. "An Optimal Batch Size for a Production System Operating Under a Just-In-Time Delivery," *International Journal of Production Economics*, 32(1993), pp. 255-260. [N]

Jarrett, G. "Logistics in the Health Care Industry," *International Journal of Physical Distribution Logistics Management*, 28 (1998), pp. 741-745. [P]

Jewkes, E., and Power, M. "A Microeconomic Analysis of Investment in Just-In-Time Manufacturing," *International Journal of Production Economics*, 29 (1993), pp. 313-321. [C]

Jina, J. "Automated JIT Based Materials Management for Lot Manufacture. *International Journal of Operations and Production Management*, 16 (1996), pp. 62-75. [C]

Johnston, R. B. "Making Manufacturing Practices Tacit: A Case of Computer-Aided Production Management and Lean Production," *Journal of the Operational Research Society*, 46 (1995), pp. 1174-1183. [C]

Johnston, S. "JIT: Maximizing its Success Potential," *Production and Inventory Management Journal*, 30 (1989), pp. 82-86. [C]

Joo, S. H., and Wilhelm, W. "A Review of Quantitative Approaches in Just-In-Time Manufacturing," *Production Planning and Control*, 4 (1993), pp. 207-222. [N, C, J]

Jordan, H. "Inventory Management and the JIT Age," *Production and Inventory Management Journal*, 29 (1988), pp. 57-60. [C]

Jordan, S. "Analysis and Approximation of a JIT Production Line," *Decision Sciences*, 19 (1988), pp. 672-681. [N, C]

Joshi, K. "Coordination in Modern and JIT Manufacturing: A Computer Based Approach," *Production and Inventory Management Journal*, 30 (1990), pp. 53-57. [C]

Joshi, K., and Campbell, J. F. "Managing Inventories in a JIT Environment," *International Journal of Purchasing and Materials Management*, 27 (1991), pp. 32-36. [C]

Jothishankar, M., and Wang, H. "Metamodelling a Just-In-Time Kanban System," *International Journal of Operations and Production Management*, 13 (1993), pp. 18-36. [K]

Junk, J. L. "Comparison of Inventory Cost Reduction Strategies in a JIT Manufacturing System," *International Journal of Production Research*, 27 (1989), pp. 1065-1090. [N, C]

Kalagana, Suresh S., and Lindsay, R. Murray "The Use of Organic Models of Control in JIT Firms: Generalizing Woodward's Findings to Modern Manufacturing Practices," *Accounting, Organizations and Society*, 24 (1999), pp. 1-4. [C, E]

Karlsson, C., and Norr, C. "Total Effectiveness in a Just-In-Time System," *International Journal of Operations and Production Management*, 14 (1994), pp. 46-65. [C]

Karmarker, U. "Getting Control of Just In Time," *Harvard Business Review*, 1989, pp. 122-133. [C]

Kazazi, A., and Keller, A. Z. "Benefits Derived from JIT by European Manufacturing Companies. (Just-In-Time)," *Industrial Management and Data Systems*, 94 (1994), pp. 12-15. [I]

Kazazi, A. "A Method for Assessing JIT Effectiveness," *Industrial Management and Data Systems*, 94 (1994), pp. 14-18. [J, N]

Kazerooni, A., Chan, F. T., and Abhary, K. "Real-Time Operation Selection in an FMS using Simulation - A Fuzzy Approach," *Production Planning and Control*, 8 (1997), pp. 771-778. [S]

Keaton, M. "A New Look at the Kanban Production Control System," *Production and Inventory Management Journal*, 36 (1995), pp. 71-78. [K]

Kelle, P., and Schneider, H. "Extension of a Reliability-Type Inventory Model to Just-In-Time Systems," *International Journal of Production Economics*, 26 (1992), pp. 319-26. [C]

Kelle, Peter, and Miller, Pam Anders "Transition to Just-In-Time Purchasing: Handling Ounce Deliveries with Vendor-purchaser Co-operation," *International Journal of Operations & Production Management*, 18 (1998), pp. 53-66. [P]

Keller, A. Z., and Kazazi, A. "Just-In-Time Manufacturing Systems: A Literature Review," *Industrial Management and Data Systems*, 93 (1993), pp. 3-22. [C]

Keller, A. Z., Kazazi, A., and Carruthers, A. "Impact of Implementing 'Just-In-Time' in a European Manufacturing Environment," *International Journal of Quality and Reliability Management*, 9 (1992), pp. 54-63. [I]

Kern, G., and Wei, J. "Master Production ReScheduling Policy in Capacity-Constrained Just-In-Time Make-To-Stock Environments," *Decision Sciences*, 27 (1996), pp. 365-387. [K]

Kevin, Z. "Tailored Just-in-Time and MRP Systems in Carpet Manufacturing," *Production & Inventory Management Journal*, 39 (1998), pp. 46-51. [P]

Kim, G. and Lee, S. M. "Impact of Computer Technology on Implementation of JIT Production Systems," *International Journal of Operations and Production Management*, 9 (1989), pp. 20-39. [C]

Kim, G. C., and Schniederjans, M. J. "An Evaluation of Computer Integrated Just In Time Production Systems," *Production and Inventory Management Journal*, 31 (1990), pp. 4-6. [E]

Kim, G. C., and Schniederjans, M. J. "A Multiple Objective Model for a Just-In-Time Manufacturing System Environment," *International Journal of Operations and Production Management*, 13 (1993), pp. 47-61. [N]

Kim, I., and Tang, C. "Lead Time and Response Time in a Pull Production Control System," *European Journal of Operational Research*, 101 (1997), pp. 474-485. [C]

Kim, T. "Just In Time Manufacturing System: A Periodic Pull System," *International Journal of Production Research*, 23 (1985), pp. 553-562. [C]

Kizilkaya, Elif, and Gupta, Surendra M. "Material Flow Control and Scheduling in a Disassemble Environment," *Computers & Industrial Engineering*, 35 (1998), pp. 93-97. [P]

Kolahan, F., and Liang, M. "An Adaptive TS Approach to JIT Sequencing with Variable Processing Times and Sequence Dependent Setups," *European Journal of Operational Research*, 109 (1998), pp. 142-148. [N]

Krajewski, L., King, B., Ritzman, L., and Wong, D. "Kanban, MRP, and Shaping the Manufacturing Environment (Material Requirements Planning)," *Management Science*, 33 (1987), pp. 39-57. [K]

Krupp, J. G. "JIT Distribution and Warehousing," *Production and Inventory Management Journal*, 30 (1991), p18-21. [P]

Kuan, K., and Changchit, C. "A Just In Time Simulation Model of a PCB Assembly Line. *International Journal of Production Research*, 20 (1991), pp. 17-26. [S]

Kubiak, W. "Minimizing Variation of Production Rates in Just-In-Time Systems: A Survey," *European Journal of Operational Research*, 66 (1993), pp. 259-271. [C]

Kubiak, W., and Sethi, S. "A Note on 'Level Schedules for Mixed-Model Assembly Lines in Just-In-Time Production Systems'," *Management Science*, 37 (1991), pp. 121-122. [N, C]

Kubiak, W., and Timkovsky, V. "Total Completion Time Minimization in Two-Machine Job Shops with Unit-Time Operations," *European Journal of Operational Research*, 94 (1996), pp. 310-320. [C]

Kung, H. K., and Changchit, C. "A Just-In-Time Simulation Model of a PCB Assembly Line," *Computers and Industrial Engineering*, 20 (1991), pp. 17-26. [C]

Lambrecht, M. R., and Decaluwe, L. "JIT and Constraint Theory: The Issue of Bottleneck Management," *Production and Inventory Management Journal*, 29 (1988), pp. 61-67. [C]

Laguna, M., and Velarde, J. "A Search Heuristic for Just-In-Time Scheduling in Parallel Machines," *Journal* of Intelligent Manufacturing, 2 (1991), pp. 253-260. [N, K]

Landry, S., Duguay, C. R., Chausse, S., and Themens, J. "Integrating MRP, Kanban and Bar-Coding Systems to Achieve JIT Procurement," *Production and Inventory Management Journal*, 38 (1997), pp. 8-12. [K]

Larson, Paul "Carrier Reduction: Impact of Logistics Performance an Interaction with EDI," *Transportation Journal*, 38 (1998), pp. 40-48. [P]

Larson, Paul D. "Air Cargo Deregulation and JIT: Ttwo 20th Anniversaries in American Logistics," *Transportation Quarterly*, 52 (1998), pp. 49-61. [K]

Larson, N., and Kusiak A. "Work-In-Process Space Allocation: A Model and an Industrial Application," *IIE Transactions*, 27 (1995), pp. 497-506. [C]

Lawley, M., Reveliotis, S., and Ferreira, P. "Flexible Manufacturing System Structural Control and the Neighborhood Policy: Part 2 - Generalization Optimization and Efficiency," *IIE Transactions*, 29 (1997), pp. 889-899. [C]

Lawrence, J., and Hottenstein, M. "The Relationship between JIT Manufacturing and Performance in Mexican Plants Affiliated with U. S. Companies," *Journal of Operations Management*, 13 (1995), pp. 3-18. [I]

Lawrence, J., and Lewis, H. "JIT Manufacturing in Mexico: Obstacles to Implementation," *Production and Inventory Management Journal*, 34 (1993), pp. 31-35. [I]

Lawrence, J., and Lewis, H. "Understanding the use of Just-In-Time Purchasing in a Developing Country: The Case of Mexico," *International Journal of Operations and Production Management*, 16 (1996), pp. 68-90. [I]

Lee, C. Y. "A Recent Development of the Integrated Manufacturing System: A Hybrid of MRP and JIT," *International Journal of Operations and Production Management*, 13 (1993), pp. 3-16. [K]

Lee, C. Y. "The Adoption of Japanese Manufacturing Management Techniques in Korean Manufacturing Industry," *International Journal of Operations and Production Management*, 12 (1992), pp. 66-81. [I]

Lee, H., and Wellan, D. "Vendor Survey Plan: A Selection Strategy for JIT/TQM Suppliers," *International Journal of Physical Distribution and Logistics Management*, 23 (1993), pp. 39-45. [Q, P]

Lee, H., and Wellan, D. "Vendor Survey Plan: A Selection for JIT/TQM Suppliers," *Industrial Management and Data Systems*, 93 (1993), pp. 8-13. [P, Q]

Lee, L. "A Comparative Study of Push and Pull Production Systems," *International Journal of Operations and Production Management*, 9 (1989), pp. 5-18. [C]

Lee, L. "Parametric Approach of the JIT System," *International Journal of Production Research*, 25 (1987), pp. 1415-1429. [N]

Lee, L., Poo, A., and Seah, K. W. "Periodic Pull: A Modified Approach to Just-In-Time Production," *International Journal of Computer Integrated Manufacturing*, 6 (1993), pp. 186-190. [C]

Lee, L., and Seah, K. "JIT and the Effects of Varying Process and Set-up Times. *International Journal of Operations and Production Management*, 8 (1988), pp. 19-35. [C]

Lee, M. K., and Kim, S. Y. "Scheduling of Storage/Retrieval Orders Under a Just-In-Time Environment," *International Journal of Production Research*, 33 (1995), pp. 3331-3348. [K]

Lee, R., and Leonard, R. "The Company-Wide Effects of Just-In-Time Manufacturing Within an Integrated Materials Handling System: A Case Study," *Production Planning and Control*, 2 (1991), pp. 364-372. [K]

Lee, S. M., and Ansari, A. "Comparative Analysis of Japanese Just In Time Purchasing and Traditional US Production Systems," *International Journal of Operations and Production Management*, 5 (1985), pp. 5-14. [P, E]

Lee, S. M., and Ansari, A. "Comparing Japanese and Traditional Purchasing in Just In Time Manufacturing," *International Trends in Manufacturing Technology*, 3 (1987), pp. 215-224. [I, E]

Lee, S. M., Chung, S., and Everett, A. "Goal Programming Methods for Implementation of Just-In-Time Production," *Production Planning and Control*, 3 (1992), pp. 175-182. [N]

Lee, S. M., and Ebrahimpour, M, "Just In Time Production System: Some Requirements for Implementation," *International Journal of Production Research*, 4 (1984), pp. 3-15. [C]

Lee, S. M., Everett, A., and Melkonian, J. "Optimizing Lot Size and Setup Time in a JIT Environment: A Multiple Objective Application," *Production Planning and Control*, 5 (1994), pp. 308-229. [N]

Lee, S. M., and Paek, J. "An Enlarged JIT Program: Its Impact on JIT Implementation and Performance of the Production System," *Production Planning and Control*, 6 (1995), pp. 185-191. [E, J]

Lee, Y. J., and Zipkin, P. "Production Control in a Kanban-like System with Defective Outputs," *International Journal of Production Economics*, 28 (1992), pp. 143-155. [K]

Lee, Y. Q., and Shin, H. J. "CIM Implementation through JIT and MRP Integration," *Computers and Industrial Engineering*, 31 (1996), pp. 609-612. [K]

Leung, F., and Kit-Leung, M. "Using Just-In-Time Analysis to Study the Maintenance Problem of Gearboxes," *International Journal of Operations and Production Management*, 16 (1996), pp. 98-105. [M]

Levasseur, G. A., and Storch, R. L. "A Non-Sequential Just-In-Time Simulation Model," *Computers and Industrial Engineering*, 30 (1996), pp. 741-752. [S]

Levner, E. V., and Nemirovsky, A. "A Network Flow Algorithm for Just-In-Time Project Scheduling," *European Journal of Operational Research*, 79 (1994), pp. 167-175. [N]

Li, S., and Qi, E. "A DEDS Model for a Serial Pull System and an Analysis of Material Movements," *International Journal of Operations and Production Management*, 15 (1995), pp. 70-88. [K]

Ling, S., and Durnota, B. "Using Two Object-Oriented Modelling Techniques: Specifying the Just-In-Time Kanban System," *Journal* of Operations and Production Management, 15 (1995), pp. 185-199. [K]

Lockamy, A., and Cox, J. "Using V-A-T Analysis for Determining Priority and Location of JIT Manufacturing Technique," *International Journal of Production Research*, 20 (1991), pp. 1661-1672. [N]

Logendran, R., and Puvanunt, V. "Duplication of Machines and Sub-Contracting of Parts in the Presence of Alternative Cell Locations," *Computers and Industrial Engineering*, 33 (1997), pp. 235-238. [C]

Lovell, M. "Simulating a 100% Just-In-Time Economy," *International Journal of Production Economics*, 26 (1992), pp. 71-78. [S]

Lowe, J., and Sim, A. B. "The Diffusion of a Manufacturing Innovation: The Case of JIT and MIRPII," *International Journal of Technology Management*, 8 (1993), pp. 244-258. [K]

Lummus, R.R. "A Simulation Analysis of Sequencing Alternatives for JIT Lines using Kanbans," *Journal of Operations Management*, 13 (1995), pp. 183-191. [S, K]

Lummus, R. R., and Duclos-Wilson, L. "When JIT is not JIT," *Production and Inventory Management Journal*, 33 (1992), pp. 61-65. [C]

Luss, H., and Rosenwen, M. B. "A Lot Sizing Model for Just In Time Manufacturing," *Journal* of the *Operations Research* Society, 41 (1990), pp. 201-209. [N]

Macbeth, D. "Supplier Management in Support of JIT activity: A Research Agenda," *International Journal of Operations and Production Management*, 7 (1987), pp. 53-63. [P]

MacMillan, J. "Principles of Point-Of-Use Storage," *Production and Inventory Management Journal*, 34 (1993), pp. 53-55. [C]

Mahmoodi, F., and Martin, G. "Optimal Supplier Delivery Scheduling to JIT Buyers," *The Logistics and Transportation Review*, 30 (1994), pp. 353-361. [P, K]

Majchrzak, A. "Just-In-Time Manufacturing: An Aggressive Manufacturing Strategy," *IEEE Transactions on Engineering Management*, 37 (1990), pp. 64-65. [C]

Malley, J., and Ray, R. "Information and Organizational Impacts of Implementing a JIT System," *Production and Inventory Management Journal*, 29 (1988), pp. 66-69. [H, C]

Manivannan, S., and Pegden, C. D. "A Rule-Based Simulator for Modelling Just-In-Time Manufacturing Systems," *IEEE Transactions Systems, Man and Cybernetics*, 20 (1990), pp. 109-117. [S]

Maniatis, P. "Technical Note: Just-In-Time in a Quality Motivated Managerial Environment," *International Journal of Materials and Product Technology*, 5 (1990), pp. 403-407. [Q]

Manoocherhri, G. "JIT for Small Manufacturers," *Journal* of Small Business Management, 26 (1988), pp. 22-30. [C]

Manoocherhri, G. "Suppliers and Just In Time Concept," *Journal* of Purchasing and Material Management, 9 (1984), pp. 16-21. [P]

Mark, Ina S., Mathieu, Richard G., and Wray, Barry A. "A Rule Induction Approach for Determining the Number of Kanbans in a Just-In-Time Production System," *Computers & Industrial Engineering*, 34 (1998), pp. 717-726. [K, N]

Markham, I., and McCart, C. D. "The Road to Successful Implementation of Just-In-Time Systems," *Production and Inventory Management Journal*, 36 (1995), pp. 67-70. [C]

Martel, M. "The Role of Just-In-Time Purchasing in Dynapert's Transition to World-Class Manufacturing," *Production and Inventory Management Journal*, 34 (1993), pp. 71-76. [P]

Marwan A.Mahmoud M. M. and Yasin, A "A Conceptual Framework for Effective Implementation JIT: An Empirical Investigation," *International Journal of Operations & Production Management*, 18 (1998), pp. 1111-1120. [E]

Mascolo, M. D. "Analysis of a Synchronization Station for the Performance Evaluation of a Kanban System with a General Arrival Process of Demands," *European Journal of Operational Research*, 89 (1996), pp. 147-163. [K]

Mascolo, M.D., Frein, Y., and Dallery, Y. "An Analytical Method for Performance Evaluation of Kanban Controlled Production Systems," *Operations Research*, 44 (1996), pp. 50-64. [K, N]

Matsuura, H., Kurosu, S., and Lehtimaki, A. "Concepts, Practices and Expectations of MRP, JIT and OPT in Finland and Japan," *International Journal of Production Economics*, 41 (1995), pp. 267-272. [K, C]

Matta, K. F. "Ordering Policies for the Cyclical Replenishment Problem Given Lead Time-Dependent Discounts," *European Journal of Operational Research*, 73 (1994), pp. 465-471. [N]

Mazany, P. "A Case Study: Lessons from the Progressive Implementation of Just-In-Time in a Small Knitwear Manufacturer," *International Journal of Operations and Production Management*, 15 (1995), pp. 271-289. [C]

Mehra, S., and Inman, R. A. "Determining the Critical Elements of Just-In-Time Implementation," *Decision Sciences*, 23 (1992), pp. 160-174. [C]

McLachlin, P. "Management Initiatives and Just-In-Time Manufacturing," *Journal of Operations Management*, 15 (1997), pp. 271-293. [C]

McTavish, R., Goyal, S., and Gunasekaran, A. "Implementation of Zero Inventories and Just-In-Time Production Concepts in Chinese Manufacturing Organizations," *Production Planning and Control*, 2 (1991), pp. 73-85. [I]

Mejabi, O., and Wasserman, G. "Basic Concepts of JIT Modelling," *International Journal of Production Research*, 30 (1992), pp. 141-149. [N, C]

Mejabi, O., and Wasserman, G. "Simulation Constructs for JIT Modeling," *International Journal of Production Research*, 33 (1992), pp. 1291-1300. [S]

Melnyk, Steven A., Calantone, Roger J., Montabon Frank L., and Smith Richard T. "Short-term Action in Pursuit of Long-term Improvement Introducing Kaizen Events," *Production & Inventory Management Journal*, 39 (1998), pp. 69-77. [C]

Meral, S., and Erkip, N. "Simulation Analysis of a JIT Production Line," *International Journal of Production Economics*, 24 (1991), pp. 147-156. [S]

Meybodi, M. Z. "Integrating Production Activity Control into a Hierarchical Production-Planning Model," *International Journal of Operations and Production Management*, 15 (1995), pp. 4-25. [N]

Millar, I. "Total Just-In-Time," *Industrial Management and Data Systems*, 2 (1990), pp. 3-10. [C]

Miller, George J. "Inventory Accuracy in 60 Days," *Hospital Materiel Management Quarterly*, 20 (1998), pp. 1-14. [C]

Miller, Pam Anders, and Kelle, Peter "Quantitative Support for Buyer-supplier Negotiation Just-In-Time Purchasing," *International Journal of Purchasing and Materials Management*, 34 (1998), pp. 25-31. [P, N]

Milliot, Jim and Mutter, John "Making More with Less: The Industry Tries to Make Just-In-Time Delivery Work," *Publishers Weekly*, 244 (1997), pp. 42-47.

Miltenburg, G. "Changing MRP's Costing Procedures to Suit JIT," *Production and Inventory Management Journal*, 31 (1990), pp. 77-83. [K, N]

Miltenburg, G., and Wijngaard, J. "The U-line Line Balancing Problem," *Management Science*, 40 (1994), pp. 1378-1388. [C]

Miltenberg, J. "Comparing JIT, MRP and TOC, and Embedding TOC into MRP," *International Journal of Production Research*, 35 (1997), pp. 1147-1169. [K]

Miltenburg, J. "Level Schedules for Mixed-Model Assembly Lines in Just-In-Time Production Systems," *Management Science*, 35 (1989), pp. 192-207. [K]

Miltenburg, J. "A Theoretical Framework for Understanding Why JIT Reduces Cost and Cycle Time and Improves Quality," *International Journal of Production Economics*, 30 (1993), pp. 195-204. [Q]

Miltenburg, J., and Sinnamon, G. Algorithms for Scheduling Multi-Level Just-In-Time Production Systems," *IIE Transactions*, 24(1992), pp. 121-130. [N, K]

Miltenburg, J., and Sinnamon, G. "Scheduling Mixed Model Multi-Level Just In Time Production System," *International Journal of Production Research*, 27 (1989), pp. 1487-1511. [K]

Miltenburg, J., and Wijngaard, J. "Designing and Phasing in Just-In-Time Production Systems," *International Journal of Production Research*, 29 (1991), pp. 115-131. [C]

Minahan, Tim "JIT: A Process with Many Faces," *Purchasing*, 123 (1997), pp. 42-47. [P]

Minahan, Tim "Transportation Key to JIT at Case Corp," *Purchasing*, 123 (1997), pp. 53-55. [P]

Ming-wei, J., and Shi-lian, L. "A Hybrid System of Manufacturing Resource Planning and Just-In-Time Manufacturing," *Computers in Industry*, 19 (1992), pp. 151-155. [C]

Mirza, M. A., and Malstrom, E. "Required Setup Reductions in JIT Driven MRP Systems," *Computers and Industrial Engineering*, 27 (1994), pp. 221-224. [K]

Mitra, D., and Mitrani, I. "Analysis of a Kanban Discipline for Cell Coordination in Production lines," *Management Science*, 36 (1990), pp. 1548-1566. [N, K]

Mittal, S., and Wang, H. "Simulation of JIT Production to Determine Number of Kanbans," *International Journal of Advanced Manufacturing Technology*, 7 (1992), pp. 292-305. [S, K]

Moeeni, F., and Chang, Y. L. "An Approximate Solution to Deterministic Kanban Systems, *Decision Sciences*, 21 (1990), pp. 596-607. [N, K]

Moeeni, F., and Chang, Y. L. "A Note on 'An Approximate Solution to Deterministic Kanban Systems': A Commentary and Further Insights," *Decision Sciences*, 27 (1996), pp. 827-832. [N, K]

Moras, R. G., and Dieck, A. "Industrial Applications of Just-In-Time: Lessons to be Learned," *Production and Inventory Management*, 33 (1992), pp. 25-29. [C]

Moras, R. G., Jalali, M. R., and Dudek, R. A. "A Categorized Survey of the JIT Literature," *Production Planning and Control*, 2 (1991), pp. 322-334. [C]

Moras, R. G., Moras, C., and Ford, R. G. "Quality Success Stories in San Antonio Industry," *Production and Inventory Management Journal*, 35 (1994), pp. 36-42. [Q]

Mould, G., and King, M. "Just-In-Time Implementation in the Scottish Electronics Industry," *Industrial Management and Data Systems*, 95 (1995), pp. 17-22. [I]

Msimangira, K. "Using "Just-In-Time" in Developing Countries: A Case Study in Tanzania," *International Journal of Purchasing and Materials Management*, 29 (1993), pp. 43-48. [I]

Muckstadt, J. A., and Tayur, S. R. "Background and Structural Results. (A Comparison of Alternative Kanban Control Mechanisms, Part 1)," *IIE Transactions*, 27 (1995), pp. 140-150. [K]

Muckstadt, J. A., and Tayur, S. R. "Experimental Results. (A Comparison of Alternative Kanban Control Mechanisms, part 2)," *IIE Transactions*, 27 (1995), pp. 151-161. [K]

Mukherjee, A., and Nof, S. Y. "Production Control for Just-In-Time Tool-and-Die Manufacturing," *Production Planning and Control*, 6 (1995), pp. 80-95. [C]

Mukhopadhyay, S. "Optimal Scheduling of Just-In-Time Purchase Deliveries," *International Journal of Operations and Production Management*, 15 (1995), pp. 59-69. [P]

Mullarkey, S., Jackson, P.R., and Parker, S. "Employee Reactions to JIT Manufacturing Practices: A Two-Phase Investigation," *International Journal of Operations and Production Management*, 15 (1995), pp. 62-79. [H]

Muralidhar, K., Swenseth, S., and Wilson, R. L. "Describing Processing Time when Simulating the JIT Environment," *International Journal of Production Research*, 30 (1992), pp. 1-11. [S]

Nakamura, Masao Sakakibara, and Schroeder, Roger "Adoption of Just-In-Time Manufacturing Methods at U. S. and Japanese-owned Plants: Some Empirical Evidence," *IEEE Transactions on Engineering Management*, 45 (1998), pp. 230-241. [I, C]

Narasimhan, R., and Melynk, S. A. "Setup Time Reduction and Capacity Management," *Production and Inventory Management Journal*, 31 (1990), pp. 55-63. [C]

Nassimbeni, G. "Factors Underlying Operational JIT Purchasing Practices: Results of an Empirical Research," *International Journal of Production Economics*, 42 (1996), pp. 275-288. [P, E]

Natarajan, R., and Goyal, S. "Safety Stocks in JIT Environments," *International Journal of Operations and Production Management*, 14 (1994), pp. 64-73. [C]

Natarajan, R., and Weinrauch, J. D. "JIT and the Marketing Interface," *Production and Inventory Management*, 3 1(1990), pp. 42-46. [C]

Neil, G., and O'Hara, J. "The Introduction of JIT into a High Technology Electronics Manufacturing Environment," *International Journal of Production Management*, 7 (1987), pp. 64-79. [C]

Nellemann, D. O., and Smith, L. F. "Just In Time vs Just In Case Production/Inventory System: Concepts Borrowed Back from Japan," *Production and Inventory Management*, 23 (1982), pp. 12-20. [C]

Nelson, P. A., and Jambeker, A. B. "A Dynamic View of Vendor Relations Under JIT," *Production and Inventory Management Journal*, 31 (1990), pp. 65-70. [P]

Neves, J. "Average Setup Cost Inventory Model: Performance and Implementation Issues," *International Journal of Production Research*, 30 (1990), pp. 445-468. [J]

Newman, R. G. "The Buyer Seller Relationships Under Just In Time," *Production and Inventory Management Journal*, 29 (1988), pp. 45-53. [P]

Newman, W., and Sridharan, V. "Manufacturing Planning and Control: Is There One Definitive Answer?," *Production and Inventory Management Journal*, 33 (1992), pp. 50-54. [C]

Ng, W., and Mak, K. L. "A Branch and Bound Algorithm for Scheduling Just-In-Time Mixed-Model Assembly Lines," *International Journal of Production Economics*, 33 (1994), pp. 169-183. [N, K]

Nori, V., and Sarker, B. R. "Cyclic Scheduling for a Multi-Product, Single-Facility Production System Operating Under a Just-In-Time Delivery policy," *Journal of the Operational Research Society*, 47 (1996), pp. 930-935. [N, K]

Norris, D. "A Study of JIT Implementation Techniques Using the Analytic Hierarchy Process Model," *Production and Inventory Management*, 33 (1992), pp. 49-53. [N]

Norris, D., Swanson, R. D., and Chu, Y. L. "Just-In-Time Production Systems: A Survey of Managers," *Production and Inventory Management Journal*, 35 (1994), pp. 63-66. [C]

Novack, R. A., Rinehart, L., and Fawcett, S. A "Rethinking Integrated Concept Foundations: A Just-In-Time Argument for Linking Production/Operations and Logistics Management," *International Journal of Operations and Production Management*, 13 (1993), pp. 31-43. [P]

O'Brien, C., and Head, M. "Developing a Full Business Environment to Support Just-In-Time Logistics," *International Journal of Production Economics*, 42 (1995), pp. 41-50. [P]

Ocana, C., and Zemel, E. "Learning From Mistakes: A Note on Just-In-Time Systems," *Operations Research*, 44 (1996), pp. 206-226. [N, J]

Occena, L. G., and Yokota, T. "Analysis of the AGV Loading Capacity in a JIT Environment," *Journal of Manufacturing Systems*, 12 (1993), pp. 24-35. [C]

Occena, L. G., and Yokota, T. "Modelling of an Automated Guided Vehicle System (AGVS) in a Just In Time Environment," *International Journal of Production Research*, 29 (1990), pp. 495-511. [C]

Offodile, O., and Arrington, D. "Support of Successful Just-In-Time Implementation: The Changing Role of Purchasing," *International Journal of Physical Distribution and Logistics Management*, 22 (1992), pp. 38-46. [P]

Oguz, C., and Dinçer, C. "Incorporating Just-In-Time into a Decision Support System Environment," *European Journal of Operational Research*, 55 (1992), pp. 344-356. [N]

Ohno, K., Nakashima, K., and Kojima, M. "Optimal Numbers of Two Kinds of Kanbans in a JIT Production System," *International Journal of Production Research*, 33 (1995), pp. 1387-1400. [K, N]

Ogan, P., and Heitger, "Alphabet Soup: Good for You or an Indigestible Stew?" *Business Horizons*, 42 (1999), pp. 61-69. [C]

Olhager, J. "Technical Note: Safety Mechanisms in Just-In-Time Systems," *International Journal of Operations and Production Management*, 15 (1995), pp. 289-292. [C]

Olhager, J., and Rapp, B. "Setup Reduction and Inventory Turnover Rate," *International Journal of Production Research*, 29 (1991), pp. 95-105. [C]

Oliver, N. "Human Factors in the Implementation of Just In Time Production," *International Journal of Operations and Production Management*, 10 (1990), pp. 32-40. [H]

Oliver, N., and Davies, A. "Adopting Japanese Style Manufacturing Methods: A Tale of Two (UK) Factories," *Journal of Management Studies*, 27 (1990), pp. 555-571. [I]

O'Neal, C. R. "The Buyer-Seller Linkage in a Just In Time Environment," Journal of Purchasing and Material Management, 30 (1989), pp. 13-30. [P]

Orth, D., Hybil, R., and Korzan, D. "Analysis of a JIT Implementation at Dover Corporation," *Production and Inventory Management*, 31 (1990), pp. 79-82. [C]

Pallares-Barbera, Montserrat "Changing Production Systems: The Automobile Industry in Spain," *Economic Geography*, 74 (1998), pp. 344-354.

Pandya, V., and Boyd, J. "Appraisal of JIT Using Financial Measures. *International Journal of Operations and Production Management*, 15 (1995), pp. 200-209. [J, C]

Park, P. "Simulation in Just-In-Time Implementation," *Simulation and Gaming*, 26 (1995), pp. 51-60. [S]

Park, S. P. "Uniform Plant Loading through Level Production," *Production and Inventory Management Journal*, 34 (1993), pp. 12-17. [C]

Parnaby, J. "A Systems Approach to the Implementation of JIT Methodologies in Lucas Industries," *International Journal of Production Research*, 26 (1988), pp. 483-492. [C]

Payne, T. "Acme Manufacturing: A Case Study in JIT Implementation," *Production and Inventory Management Journal*, 34 (1993), pp. 82-86. [C]

Pegler, H., and Kochhar, A. "Rule-Based Approach to Just-In-Time Manufacturing," *Computer-Integrated Manufacturing Systems*, 3 (1990), pp. 11-19. [N]

Perkins, J. R. "Hedging Policies for Failure-Prone Manufacturing Systems: Optimality of JIT Bound Buffer Levels," *IEEE Transactions on Automatic Control*, 43 (1998), pp. 953-959. [N]

Perkins, J., and Kumar, P. R. "Optimal Control of Pull Manufacturing Systems," *IEEE Transactions on Automatic Control*, 40 (1995), pp. 2040-2051. [C]

Perry, J. "Firm Behavior and Operating Performance in Just In Time Logistics Channels," *Journal of Business Logistics*, 9 (1988), pp. 19-42. [P]

Peters, M., and Austin, M. "The Impact of JIT: A Critical Analysis," *Industrial Management and Data Systems*, 95 (1995), pp. 12-17. [C]

Philipoom, P. R., Rees, L., and Taylor, B. W. "Simultaneously Determining the Number of Kanbans, Container Sizes, and the Final Assembly Sequence of Products in a Just-In-Time Shop," *International Journal of Production Research*, 34 (1996), pp. 51-69. [K]

Philipoom, P. R., Rees, L., Taylor, B. W., and Hung, P. Y. "A Mathematical Programming Approach for Determining Workcenter Lotsizes in a Just In Time System with Signal Kanbans," *International Journal of Production Research*, 28 (1990), pp. 1-17. [N, K]

Plenert, G. "An Overview of JIT," *International Journal of Advanced Manufacturing Technology*, 8 (1993), pp. 91-95. [C]

Plenert, G. "Are Japanese Production Methods Applicable in the United States," *Production and Inventory Management Journal*, 26 (1985), pp. 119-129. [C]

Plenert, G. "Line Balancing Techniques as Used for Just-In-Time (JIT) Product Line Optimization," *Production Planning and Control*, 8 (1997), pp. 686-693. [N]

Plenert, G. "Three Differing Concepts of JIT," *Production and Inventory Management Journal*, 31 (1990), pp. 1-2. [C]

Plenert, G., and Best, T.D. "MRP, JIT and OPT: What's Best?," *Production and Inventory Management Journal*, 27 (1986), pp. 22-29. [K]

Pleschberger, T., and Hitomi, K. "Flexible Final-Assembly Sequencing Method for a JIT Manufacturing Environment," *International Journal of Production Research*, 31 (1993), pp. 1189-1199. [C]

Pleschberger, T., and Hitomi, K. "Just-In-Time Shipments in a Truck-Traffic-Coordination System," *International Journal of Production Economics*, 33 (1994), pp. 195-205. [P]

Porter, Anne Millen "The Problem with JIT," *Purchasing*, 123 (1997), pp. 18-20. [P]

Pourbabai, B. "Loading Strategies for a Class of Just-In-Time Manufacturing System," *International Journal of Advanced Manufacturing Technology*, 10 (1995), pp. 46-51. [K]

Pourbabai, B. "Optimal Selection of Order in a Just In Time Manufacturing Environment: A Loading Model for a Computer Integrated Manufacturing System," *International Journal of Computer Integrated Manufacturing*, 5 (1992), pp. 38-44. [N]

Power, Damien J., and Sohal, Amrik S. "An Examination of the Literature Relating to Issues Affecting the Human Variable in Just-In-Time Environment," *Technovation*, 17 (1997), pp. 649-667. [H]

Prasad, B. "JIT Quality Matrices for Strategic Planning and Implementation," *International Journal of Operations and Production Management*, 15 (1995), pp. 116-142. [Q]

Price, W., Gravel, M., and Nsakanda, A. L. "A Review of Optimization Models of Kanban-Based Production Systems," *European Journal of Operational Research*, 75 (1994), pp. 1-12. [N, K]

Primrose, P. L. "Evaluating the Introduction of JIT," *International Journal of Production Economics*, 27 (1992), pp. 9-22. [C]

Ptak, C.A. "MRP, MRP II, OPT, JIT and CIM-Succession Evaluation, or Necessary Combination," *Production and Inventory Management Journal*, 32 (1991), pp. 7-11. [S]

Pun, K. F, Chin, K. S., and Wong, K. "Implementing JIT in a PCB Manufacturer," *Production and Inventory Management Journal*, 39 (1998), pp. 10-17. [C]

Pyke, D. F., and Cohen, M. A. "Push and Pull in Manufacturing and Distribution Systems," *Journal of Operations Management*, 9 (1990), pp. 24-33. [C]

Rajendran, C., and Ziegler, H. "An Efficient Heuristic for Scheduling in a Flow Shop to Minimize Total Weighted Flow Time of Jobs," *European Journal of Operational Research*, 103 (1997), pp. 129-138. [N, S]

Ramani, S., and Narayanan, N. "Single Facility, Multi-Item Lot Sizing Under Just-In-Time and "Cyclic Scheduling for Improvement," *International Journal of Production Economics*, 26 (1992), pp. 333-339. [N, K]

Ramarapu, N., Mehra, S., and Frolick, M. "A Comparative Analysis and Review of JIT "Implementation" Research," *International Journal of Operations and Production Management*, 15 (1995), pp. 38-49. [C]

Ramasesh, R. V. "A Logistics-Based Inventory Model for JIT Procurement," *International Journal of Operations and Production Management*, 13 (1993), pp. 44-58. [P]

Ramesesh, R. V. "Recasting the Traditional Inventory Model to Implement Just-In-Time Purchasing," *Production and Inventory Management*, 31 (1991), pp. 71-75. [C]

Rao, A. "Manufacturing Systems - Changing to Support JIT," *Production and Inventory Management*, 30 (1989), pp. 18-21. [C]

Rao, A. "Moving From Manufacturing Resources Planning to Just In Time Manufacturing," *Production and Inventory Management Journal*, 29 (1988), pp. 44-49. [C]

Rao, A. "A Survey of MRPII Software Suppliers: Trends in Support of Just In Time," *Production and Inventory Management Journal*, 30 (1989), pp. 14-17. [K]

Rees, L., Huang, P. Y., and Taylor, B. W. "Comparative Analysis of a MRP Lot-For-Lot System and a Kanban System for a Multistage Production Operation," *International Journal of Production Research*, 27 (1989), pp. 1427-1443. [K]

Rees, L., Philipoom, P. R., and Huang, P.Y. "Dramatically Adjusting the Number of Kanbans in a Just In Time Production System Using Estimated Values of Lead Time," *IIE Transactions*, 19 (1987), pp. 199-207. [K, N]

Richeson, L., Lackey, C. W., and Starner Jr., J. W. "The Effect of Communication on the Linkage Between Manufacturers and Suppliers in a Just-In-Time Environment," *International Journal of Purchasing and Materials Management*, 31 (1995), pp. 21-28. [P]

Richmond, I., and Blackstone, J. "Just In Time in the Plastics Industry," *International Journal of Production Research*, 26 (1988), pp. 27-34. [C]

Ritzman, L., King, B., and Krajewski, J. H. "Manufacturing Performance - Pulling the Right Levers," *Harvard Business Review*, 62 (1984), pp. 143-152. [C]

Roach, A., and Nagi, R. "A Hybrid GA-SA Algorithm for Just-In-Time Scheduling of Multi-Level Assemblies," *Computers and Industrial Engineering*, 30 (1996), pp. 1047-1060. [N, K]

Rohleder, T. R., and Scudder, G. D. "An Experimental Comparison of Time-Based and Economic-Based Scheduling Methods to Maximize Net Present Value," *Decision Sciences*, 24 (1993), pp. 1037-1056. [K, N]

Roman, Leigh Ann "Just-In-Time: Methodist Saves Big with More Deliveries," *Memphis Business Journal*, 20 (1999), pp. 3-6. [C, P]

Romero, B. "The Other Side of JIT in Supply Management," *Production and Inventory Management Journal*, 32 (1991), pp. 1-2. [P]

Ronen, B., and Karp, R. "An Information Entropy Approach to the Small-Lot Concept," *IEEE Transactions on Engineering Management*, 41 (1994), pp. 89-92. [C]

Rosenberg, L., and Cambell, D. "Just In Time Inventory Control: A Subset of Channel Management," *Journal of the Academy of Marketing Science*, 13 (1985), pp. 124-133. [C]

Rosenblatt, Meir J., Hefter, Ilan, and Herer, Yale T. "Note: An Acquisition Policy for a Single Item Multi-supplier System," *Management Science*, 44 (1998), pp. 96-97. [P]

Roy, R. N., and Guin, K. K. "A Proposed Model of JIT Purchasing in an Integrated Steel Plant," *International Journal of Production Economics*, 59 (1999), pp. 179-188. [P]

Safayeni, F., and Purdy, L. "A Behavioral Case Study of Just-In-Time Implementation," *Journal of Operations Management*, 10 (1991), pp. 213-228. [H]

Safayeni, F., Purdy, L., Englen, R. V., and Pal, S. "Difficulties of Just In Time Implementation: A Classification Scheme," *International Journal of Operations and Production Management*, 11 (1991), pp. 27-36. [C]

Sakakibara, S., and Flynn, B. B. "The Impact of Just-In-Time Manufacturing and its Infrastructure on Manufacturing Performance," *Management Science*, 4 (1997), pp. 1246-1257. [N, C]

Samaddar, S., and Kaul, T. "Effects of Setup and Processing Time Reductions on WIP in the JIT Production Systems (Work-In-Process; Just-In-Time)," *Management Science*, 41 (1995), pp. 1263-1265. [C]

Santacecilia, P. "Increasing Manufacturing Competitiveness Through Information Technology: A Case Study," *Production and Inventory Management Journal*, 33 (1992), pp. 80-84. [C]

Sargent, T. A., and Kay, M. G. "Implementation and Utilization of a Decentralized Storage System: Costing Model," *International Journal of Operations and Production Management*, 15 (1995), pp. 210-219. [N, C]

Sarker, B. R. "Service Time Distributions and the Performance of a Pull System: A Simulation Study," *Production Planning and Control*, 2 (1991), pp. 36-43. [S]

Sarker, B. R. "Simulating a Just In Time Production System," *Computers and Industrial Engineering*, 16 (1989), pp. 127-137. [S]

Sarker, B. R., and Balan, C. V. "Operations Planning for Kanbans Between Two Adjacent Workstations," *Computers and Industrial Engineering*, 31 (1996), pp. 221-224. [K]

Sarker, B. R., and Fitzsimmons, J. A. "The Performance of Push and Pull Systems: A Simulation and Comparative Study," *International Journal of Production Research*, 27 (1989), pp. 1715-1731. [S]

Sarker, B. R., and Harris, R. D. "Effect of Imbalance in a Just In Time Production System: A Simulation Study," *International Journal of Production Research*, 26 (1988), pp. 1-18. [S]

Sarker, R., and V. Balan, Chidambaram "Operations Planning for a Multi-Stage Kanban System," *European Journal of Operational Research*, 112 (1999), pp. 284-286. [K]

Savsar, M. "Effects of Kanban Withdrawal Policies and Other Factors on the Performance of JIT Systems -- A Simulation Study," *International Journal of Production Research*, 34 (1996), pp. 2879-2899. [K, S]

Savsar, M. "Simulation Analysis of Maintenance Policies in Just-In-Time Production Systems," *International Journal of Operations and Production Management*, 17 (1997), pp. 256-266. [M, S]

Savsar, M., and Al-Jawini, A. "Simulation Analysis of Just-In-Time Production Systems," *International Journal of Production Economics*, 42 (1995), pp. 67-78. [S]

Savsar, Mehmet "Simulation Analysis of a Pull-push System for an Electronic Assembly Line," *International Journal of Production Economics*, 51 (1997), pp. 205-215. [S, K]

Schmenner, Roger W. "The Merit of Making Things *Fast,*" *Sloan Management Review*, 30 (1988), pp. 11-17. [C]

Schneider, J. D., and Leatherman, M. A. "Integrated Just-In-Time: A Total Business Approach," *Production and Inventory Management Journal*, 33 (1992), pp. 78-82. [C]

Schniederjans, Marc J. and Qing, Cao "A Note on An Extention of JIT Purchasing vs. EOQ with a Price Discount: An Analytical Comparison of Inventory Costs," *International Journal of Production Economics*, (in press). [J, N]

Schoenberger, R. J. "Applications of Single Card and Dual Card Kanban," *Interfaces*, 13 (1983), pp. 56-67. [K]

Schoenberger, R. J. "Selecting the Right Manufacturing Inventory System: Western and Japanese Approaches," *Production and Inventory Management*, 24 (1983), pp. 33-44. [C]

Schoenberger, R. J. "Some Observations on the Advantages and Implementation Issues of Just In Time Productions Systems," *Journal of Operations Management*, 3 (1982), pp. 1-12. [C]

Schoenberger, R. J., and Ansari, A. "Just In Time Purchasing Can Improve Quality," *Journal of Purchasing and Materials Management*, 20 (1986), pp. 2-7. [Q]

Schoenberger, R. J., and Gilbert, J. "Just In Time Purchasing: A Challenge for US Industry," *California Management Review*, 26 (1983), pp. 54-68. [P]

Schoenberger, R. J., and Schniederjans, M. J. "Reinventing Inventory Control," *Interfaces*, 14 (1984), pp. 76-83. [C, N]

Schole, M. and Fish, A. "Just-In-Time Inventories in Old Detroit," *Business History*, 40 (1998), pp. 48-72. [C]

Schultz, Kenneth L., Jura, David C., Boudreau, John W., McClain, John O., and Thomas, L. Joseph "Modeling and Worker Motivation in JIT Production Systems," *Management Science*, 44 (1998), pp. 1595-1605. [N, H]

Scott, A. F., Macomber, J., and Ettkin, L. "JIT and Job Satisfaction: Some Empirical Results," *Production and Inventory Management*, 33 (1992), pp. 36-41. [H, E]

Seidman, T., and Humes Jr., C. "Some Kanban-Controlled Manufacturing Systems: A First Stability Analysis," *IEEE Transactions on Automatic Control*, 41 (1996), pp. 1013-1018. [K]

Seki, Y., and Naoto H. "Transient Behavior of a Single-Stage Kanban System Based on the Queuing Model," *International Journal of Production Economics*, 60 (1999), pp. 369-374. [K, N]

Sevier, A. "Managing Employee Resistance to JIT: Creating an Atmosphere that Facilitates Implementation," *Production and Inventory Management Journal*, 33 (1992), pp. 83-87. [H]

Sewell, G. "Management Information Systems for JIT Production," *Omega*, 18 (1990), pp. 491-503. [C]

Sharadapriyadarshini, B., and Rajendran, Chandrasekharan "Heuristics for Scheduling in a Kanban System with Dual Blocking Mechanisms," *European Journal of Operational Research*, 103 (1997), pp. 439-453. [K, N]

Sherrard, W. R., and Hampton, D. R. "Passing the APICS Certification Exams: Is Thinking Required?," *Production and Inventory Management Journal*, 39 (1998), pp. 20-24. [C]

Shin, D., and Min, H. "An Analysis of Line-Stop Strategy in Just-In-Time Manufacturing," *International Journal of Operations and Production Management*, 15 (1995), pp. 104-115. [C]

Shin, D., and Min, H. "A Line-Stop Strategy for the Just-In-Time/Total Quality control Production System," *Production and Inventory Management Journal*, 34 (1993), pp. 43-47. [Q]

Shin, D., and Min, H. "Flexible Line Balancing Practices in a Just In Time Environment," *Production and Inventory Management Journal*, 32 (1991), pp. 38-41. [C]

Shin, D., and Min, H. "Uniform Assembly Line Balancing with Stochastic Times in Just In Time Manufacturing," *International Journal of Operations and Production Management*, 11 (1991), pp. 23-34. [N]

Siha, S. "Modeling the Blocking Phenomenon in JIT Environment: An Alternative Scenario," *Computers and Industrial Engineering*, 30 (1996), pp. 61-75. [N]

Siha, S., and David, H. T. "A Finch-Foster Duality Between 'Pull' and 'Push' Production Systems," *Journal of the Operational Research Society*, 45 (1994), pp. 179-186. [C]

Sillince, J., and Sykes, G. "Integrating MRPII and JIT: A Management Rather than a Technical Challenge," *International Journal of Operations and Production Management*, 13 (1993), pp. 18-31. [K]

Silver, E. "Changing the Givens in Modeling Inventory Problems: The Example of Just-In-Time Systems," *International Journal of Production Economics*, 26 (1992), pp. 347-351. [N]

Silver, E. "Joint Selection of the Purchase Order Quantity and the Number of Deliveries in a Quantity Discount/JIT Environment," *Production and Inventory Management Journal*, 36 (1995), pp. 82-83. [N]

Simpson, M, Sykes, G., and Abdullah, A. "Case Study: Transitory JIT at Proton cars, Malaysia," *International Journal of Physical Distribution and Logistics Management*, 28 (1998), pp. 121-125. [I]

Singh, N., and Brar, J. "Modelling and Analysis of Just In Time Manufacturing Systems: A Review," *International Journal of Operations and Production Management*, 12 (1992), pp. 3-14. [N, C]

Sipper, D. "JIT vs WIP - A Trade Off Analysis," *International Journal of Production Research*, 27 (1989), pp. 903-916. [C]

Slack, N., and Correa, H. "The Flexibilities of Push and Pull," *International Journal of Operations and Production Management*, 12 (1992), pp. 82-92. [C]

Spence, A., and Porteus, E. L. "Setup Reduction and Increased Effective Capacity," *Management Science*, 33 (1987), pp. 1291-1301. [N]

Spencer, M. S. "Cycle Counting in a JIT Environment Using V-A-T Focusing," *International Journal of Production Research*, 33 (1995), pp. 1699-1708. [C]

Spencer, M. S. "The Impact of JIT on Capacity Management: A Case Study and Analysis," *Production Planning and Control*, 8 (1997), pp. 183-193. [C]

Spencer, M. S., Daugherty, P., and Rogers, D. S. "Logistics Support for JIT Implementation," *International Journal of Production Research*, 34 (1996), pp. 701-714. [P]

Spencer, M. S., and Guide, V. D. "An Exploration of the Components of JIT: A Case Study and Survey Results," *International Journal of Operations and Production Management*, 15 (1995), pp. 72-83. [C]

Spencer, M. S., Rogers, D. S., and Daugherty, P. "JIT Systems and External Logistics Suppliers," *International Journal of Operations and Production Management*, 14 (1994), pp. 60-74. [P]

Smolowitz, I. "JIT - A Revolution in Viewing Business School Curricula: A Note," *Production and Inventory Management Journal*, 33 (1992), pp. 87-88. [C]

Sohal, A., Keller, A .Z., and Fouad, R. "Review of Literature Relating to JIT," *International Journal of Operations and Production Management*, 9 (1989), pp. 15-25. [C]

Sohal, A., Lewis, G., and Samson, D. "Integrating CNC Technology and the JIT Kanban System: A Case Study," *International Journal of Technology Management*, 8 (1993), pp. 422-431. [K]

Sohal, A., and Naylor, D. "Implementation of JIT in a Small Manufacturing Firm," *Production and Inventory Management Journal*, 33 (1992), pp. 20-26. [C]

Sohal, A., Ramsay, L., and Samson, D. "JIT Manufacturing: Industry Analysis and a Methodology for Implementation," *International Journal of Physical Distribution and Logistics Management*, 23 (1993), pp. 4-21. [C]

South, J. B. "A Modified Standard Cost-Accounting System Can Generate Valid Product Costs," *Production and Inventory Management Journal*, 34 (1993), pp. 28-31. [N]

Spearman, M. L. "Customer Service in Pull Production Systems," *Operations Research*, 40 (1992), pp. 948-958. [C]

Spencer, M. S. "Production Planning in a MRP/JIT Repetitive Manufacturing Environment," *Production Planning and Control*, 6 (1995), pp. 176-184. [K]

Spencer, M. S., Rogers, D. S., and Daugherty, P. J. "JIT Systems and External Logistics Suppliers," *International Journal of Operations and Production Management*, 14 (1994), pp. 60-74. [P]

Spencer, M. S., Daugherty, P. J., and Rogers, D. S. "Towards a Deeper Understanding of JIT: a Comparison Between APICS and Logistics Managers," *Production and Inventory Management*, 35 (1994), pp. 23-28. [P]

Spencer, M. S., and Guide, V. D. "An Exploration of the Components of JIT: Case Study and Survey Results," *International Journal of Operations and Production Management*, 15 (1995), pp. 72-83. [C]

Srinivasan, K., Kekre, S., and Mukhopadhyay, T. "Impact of Electronic Data Interchange Technology on JIT Shipments," *Management Science*, 40 (1994), pp. 1291-1304. [N, C]

Sriparavastu, L., and Gupta, T. "An Empirical Study of Just-In-Time and Total Quality Management Principles Implementation in Manufacturing firms in the USA," *International Journal of Operations and Production Management*, 17 (1997), pp. 1215-1232. [E, Q]

Stahl. Robert "Cycle Counting: A Quality Assurance Process," *Hospital Materiel Management Quarterly*, 20 (1998), pp. 22-29. [C, Q]

Steiner, G., and Truscott, W. G. "Batch Scheduling to Minimize Cycle Time, Flow Time, and Processing Cost," *IIE Transactions*, 25 (1993), pp. 90-96. [C]

Steiner, G., and Yeomans, J. "Level Schedules for Mixed-Model, Just-In-Time Processes," *Management Science*, 39 (1993), pp. 728-735. [K]

Steiner, G., and Yeomans, J. "Optimal Level Schedules in Mixed-Model, Multi-Level JIT Assembly Systems With Pegging," *European Journal of Operational Research*, 95 (1996), pp. 38-52. [K]

Stock, James R. "Just-In-Time for the 90s: A Wholesaler-Distributor's Guide to JIT Inventory Management," *International Journal of Physical Distribution and Logistics Management*, 24 (1994), pp. 43-54. [C]

Stockton, D., and Lindley, R. "Implementing Kanbans Within High Variety/Low Volume Manufacturing Environments," *International Journal of Operations and Production Management*, 15 (1995), pp. 47-59. [K]

Storhagen, N. G. "The Human Aspects of JIT Implementation," *International Journal of Physical Distribution and Logistics Management*, 25 (1995), pp. 4-23. [H]

Storhagen, N. G., and Hellberg, R. "Just In Time from a Business Logistics Perspective," *Engineering Costs and Production Economics*, 12 (1987), pp. 117-121. [P]

Su, K.D. "How a Leading Heavy Industries Co., Ltd. in Korea Implements JIT Philosophy to Its Operations," *Computers and Industrial Engineering*, 27 (1994), pp. 5-9. [I]

Sule, D. R., and Norris, W. "Manpower Assignment Strategies in Serial Production Lines with Limited Personnel: Evaluation of Pull-Push Rules," *Computers and Industrial Engineering*, 22 (1992), pp. 231-243. [H]

Sumichrast, R. T., and Clayton, E. R. "Evaluating Sequences for Paced, Mixed-Model Assembly Lines with JIT Component Fabrication," *International Journal of Production Research*, 34 (1996), pp. 3125-3143. [C]

Sumichrast, R. T., and Russell, R. "Evaluating Mixed-Model Assembly Line Sequencing Heuristics for Just-In-Time Production Systems," *Journal of Operations Management*, 9 (1990), pp. 371-390. [N]

Sumichrast, R. T., Russell, R., and Taylor, B. W. "A Competitive Analysis of Sequencing Procedures for Mixed Model Assembly Lines in a Just In Time Production System," *International Journal of Production Research*, 30 (1992), pp. 199-214. [C]

Suri, R., and DeTreville, S. "Getting From Just In Case to Just In Time: Insights From a Simple Model," *Journal of Operations Management*, 6 (1986), pp. 295-304. [C]

Swenseth, S. R., and Buffa, F. "Implications of Inbound Lead variability for Just In Time Manufacturing," *International Journal of Operations and Production Management*, 11 (1991), pp. 37-46. [C]

Swenseth, S. R., Muralidhar, K., and Wilson, R. L. "Planning for Continual Improvement in a Just-In-Time Environment," *International Journal of Operations and Production Management*, 13 (1993), pp. 4-22. [C]

Takahashi, K. "Determining the Number of Kanbans for Unbalanced Serial Production Systems," *Computers and Industrial Engineering*, 27 (1994), pp. 213-216. [K]

Takahashi, K., Nakamura, N., and Izumi, M. "Concurrent Ordering in JIT Production Systems," *International Journal of Operations and Production Management*, 17 (1997), pp. 267-290. [C, P]

Tatikonda, M. V. "Just In Time and Modern Manufacturing Environments: Implications for Cost Accounting," *Production and Inventory Management Journal*, 29 (1988), pp. 1-5. [C]

Temponi, C., and Pandy, S. Y. "Implementation of Two JIT Elements in Small-Sized Manufacturing Firms," *Production and Inventory Management Journal*, 36 (1995), pp. 23-28. [C]

Thomas, P. "Kanban: Just-In-Time at Toyota: Management Begins at the Workplace," *Interfaces*, 19 (1989), pp. 113-115. [K]

Toomey, J. W. "Establishing Inventory Control Options for Just In Time Application," *Production and Inventory Management Journal*, 29 (1989), pp. 13-15. [C]

Trelevein, M. "The Timing of Labor Transfers in Dual Resource-Constrained Systems: 'Push' vs. 'Pull' Rules," *Decision Sciences*, 18 (1987), pp. 73-88. [N]

Trentesaux, D., Tahon, C., and Ladet, P. "Hybrid Production Control Approach for JIT Scheduling," *Artificial Intelligence in Engineering*, 12 (1998), pp. 49-67. [K, N]

Tsiushuang, C., Xiangtong , Q., and Fengsheng, T. "Single Machine Scheduling to Minimize Weighted Earliness Subject to Maximum Tardiness," *Computers and Operations Research*, 24 (1997), pp. 147-152. [K, N]

Turnbull, P., Oliver, N., and Wilkinson, B. "Buyer-Supplier Relations in the UK Automotive Industry: Strategic Implications of the Japanese Manufacturing Model," *Strategic Management Journal*, 13 (1992), pp. 159-168. [P, I]

Upton, Da "Just-In-Time and Performance Measurement Systems," *International Journal of Operations & Production Management*, 18 (1998), pp. 1101-1110. [C, J]

Vastag, G., and Whybark, D. C. "Global Relations Between Inventory, Manufacturing Lead Time and Delivery Date Promises," *International Journal of Production Economics*, 30 (1993), pp. 563-570. [C]

Vembu, S., and Srinivasan, G. "Heuristics for Operator Allocation and Sequencing in Just-In-Time Flow Line Manufacturing Cell," *Computers and Industrial Engineering*, 29 (1995), pp. 309-313. [N]

Vemuganti, G., Batta, R., and Zhu, Y. "A Note on 'An Approximate Solution to Deterministic Kanban Systems," *Decision Sciences*, 27 (1996), pp. 817-816. [K, N]

Vendemia, W. G., Patuwo, B. E., and Hung, M. S. "Evaluation of Lead Time in Production/Inventory Systems with Non-Stationary Stochastic Demand," *Journal of the Operational Research Society*, 46 (1995), pp. 221-233. [N]

Vickery, S. K. "International Sourcing: Implications for Just In Time Manufacturing," *Production and Inventory Management Journal*, 29 (1989), pp. 66-73. [P]

Villeda, R., Dudek, R., and Smith, M. L. "Increasing the Production Rate of a Just In Time Manufacturing System with Variable Operation Times," *International Journal of Production Research*, 26 (1987), pp. 1749-1768. [C]

Vokurka, R. J., and Davis, R. A. "Just-In-Time: The Evolution of a Philosophy," *Production and Inventory Management Journal*, 37 (1996), pp. 56-59. [C]

Vondermebse, M., Tracey, M., Tan, C. L., and Bardi, Edward J. "Current Purchasing Practices and JIT: Some of the Effects on Inbound Logistics," *International Journal of Physical Distribution and Logistics Management*, 25 (1995), pp. 33-48. [P]

Vora, J. A. "Applying a Theory of Organization Change to JIT, Just-In-Time Manufacturing," *Omega*, 20 (1992), pp. 193-199. [H]

Vora, J. A., Saraph, J. V., and Peterson, D. L. "JIT Implementation Practices," *Production and Inventory Management*, 31 (1990), pp. 57-59. [C]

Voss, C. A. "Japanese Manufacturing Management Techniques in the UK," *International Journal of Operations and Production Management*, 14 (1983), pp. 31-38. [I]

Voss, C. A., and Robinson, S. "Application of Just In Time Manufacturing Techniques in the United Kingdom," *International Journal of Operations and Production Management*, 7 (1987), pp. 46-52. [I]

Vuppalapati, K., Ahire, S. L., and Gupta, T. "JIT and TQM: A Case for Joint Implementation," *International Journal of Operations and Production Management*, 15 (1995), pp. 84-94. [Q]

Wacker, J. G. "Can Holding Cost be Overstated for Just In Time Manufacturing Systems?," *Production and Inventory Management*, 27 (1986), pp. 11-14. [C]

Wafa, M. A., and Yasin, M. "The Effect of Situational Constraints on Workforce Performance and JIT Implementation: An Empirical Study," *International Journal of Computer Applications in Technology*, 8 (1995), pp. 139-144. [E, H]

Wallace, Bob "Just-In-Time Technology Put the Brakes on GM Line Productivity," *Computerworld*, 32 (1998), pp. 6-8.

Walleigh, R. "What is Your Excuse for Not Using JIT," *Harvard Business Review*, 64 (1986), pp. 38-54. [C]

Wang, D., "Earliness/Tardiness Production Planning Approaches for Manufacturing Systems," *Computers and Industrial Engineering*, 28 (1995), pp. 425-436. [C]

Wang, H., and Wang, H. P. "Optimum Number of Kanbans Between Two Adjacent Workstations in a JIT System," *International Journal of Production Economics*, 22 (1991), pp. 179-180. [N, K]

Wang, H. P., Zeng, L., and Jin, S. "Determination of the Number of Kanbans in JIT Systems: A Petri Net Approach," *International Journal of Advanced Manufacturing Technology*, 7 (1992), pp. 51-57. [N, K]

Wang, Wei, Wang, Dingwei, and Ip, W. H. "JIT Production Planning Approach with Fuzzy Due Date OKP Manufacturing Systems," *International Journal of Production Economics*, 58 (1999), pp. 20-31. [K, N]

Waples, E., and Norris, D. M. "Just In Time Production and the Financial Audit," *Production and Inventory Management Journal*, 29 (1989), pp. 25-27. [J]

Wasco, W., Stonehocker, R. E., and Feldman, L. "Success With JIT and MRP II in a Service Organization," *Production and Inventory Management Journal*, 32 (1991), pp. 16-21. [K]

Watanabe, N., and Hiraki, S. "An Approximate Solution to a JIT-Based Ordering System. *Computers and Industrial Engineering*, 31 (1996), pp. 565-569. [P]

Waters-Fuller, N. "Just-In-Time Purchasing and Supply: A Review of the Literature," *International Journal of Operations and Production Management*, 15 (1995), pp. 220-236. [P]

Weber, C. A., and Ellram, L. "Supplier Selection Using Multi-Objective Programming: A Decision Support System Approach," *International Journal of Physical Distribution and Management*, 23 (1993), pp. 3-14. [N]

Welgama, P., and Mills, R.G. "Use of Simulation in the Design of a JIT System," *International Journal of Operations and Production Management*, 15 (1995), pp. 245-260. [S]

Weng, Z. K. "Tailored Just-In-Time and MRP Systems in Carpet Manufacturing," *Production and Inventory Management Journal*, 39 (1998), pp. 46-50. [C]

Westbrook, R. "Time to Forget Just In Time? Observations on Visit to Japan," *International Journal of Operations and Production Management*, 8 (1988), pp. 5-21. [C]

White, M. "The Japanese Style of Management in Britain," *International Journal of Operations and Production Management*, 3 (1983), pp. 14-21. [I]

White, R. "An Empirical Assessment of JIT in US Manufacturers," *Production and Inventory Management Journal*, 34 (1993), pp. 38-42. [E]

White, R. E., Pearson, J. N., and Wilson, Jeffrey R. "JIT Manufacturing: A Survey of Implementations in Small and Large U.S. Manufacturers," *Management Science*, 45 (1999), pp. 1-15. [E]

Whitson, Daniel. "Applying Just-In-Time Systems in Health Care," *IIE Solutions*, 29 (1997), pp. 32-38. [C]

Wilamowsky, Y., Epstein, S., and Dickman, B. "Optimal Common Due-Date Completion Time Tolerance," *Computers and Operations Research*, 23 (1996), pp. 1203-1210. [C]

Wildemann, H. "Implementation Strategies for the Integration of Japanese Kanban - Principles in German Companies," *Engineering and Production Economics*, 9 (1985), pp. 305-319. [C, K]

Wildemann, H. "Just In Time Production in West Germany," *International Journal of Production Research*, 26 (1988), pp. 521-535. [I]

Wilkinson, B., and Oliver, N. "Power, Control and Kanban," *Journal of Management Studies*, 26 (1989), pp. 47-58. [K]

Williams, W., and Haslam, C. "Why Talk the Stock Out? Britain vs. Japan," *International Journal of Operations and Production Management*, 9 (1988), pp. 91-105. [I, E]

Willis, T., and Huston, R. "Vendor Requirement and Evaluation in a Just In Time Environment," *International Journal of Operations and Production Management*, 10 (1990), pp. 106-121. [P]

Willis, T., Huston, R., and Pohlkamp, F. "Evaluation Measures of Just-In-Time Supplier Performance," *Production and Inventory Management Journal*, 34 (1993), pp. 1-5. [P]

Willis, T., and Suter, W. C. "The Five M's of Manufacturing: A JIT Conversion Life Cycle," *Production and Inventory Management Journal*, 30 (1989), pp. 53-57. [C]

Wilson, J. "Henry Ford: A Just-In-Time Pioneer," *Production and Inventory Management Journal*, 37 (1996), pp. 26-31. [C]

Wilson, J. "Henry Ford's Just-In-Time System," *International Journal of Operations and Production Management*, 15 (1995), pp. 59-75. [C]

Wilson, James M. "A Comparison of the 'American System of Manufacturer circa 1850 with Just in Time Methods," *Journal of Operations Management*, 16 (1998), pp. 77-91. [C]

Wisner, J. D. "A Study of US Machine Shops With Just-In-Time Customers," *International Journal of Operations and Production Management*, 16 (1996), pp. 62-76. [C]

Withers, B., Ebrahimpour, M., and Hikment, N. "An Exploration of the Impact of TQM and JIT on ISO 9000 Registered Companies," *International Journal of Production Economics*, 53 (1997), pp. 209-216. [Q]

Woolsey, R., Bowden, R. O., Hall, J. D. and Hall, W. H. "Closed-Form Solution to Kanban Sizing Problem," *Production & Inventory Management Journal*, 40 (1999), pp. 1-4. [K]

Wouters, M. F. "Economic Evaluation of Lead Time Reduction," *International Journal of Production Economics*, 22 (1991), pp. 111-120. [J]

Wray, B. A., Rakes, T. R., and Rees, L. "Neural Network Identification of Critical Factors in a Dynamic Just-In-Time Kanban Environment," *Journal of Intelligent Manufacturing*, 8 (1997), pp. 83-96. [N, K]

Xiaobo, Z., and Ohno, K. "Algorithms for Sequencing Mixed Models on an Assembly Line in a JIT Production System," *Computers and Industrial Engineering*, 32 (1997), pp. 47-56. [C]

Xiaobo, Z., and Ohno, K. "A Sequencing Problem for a Mixed-Model Assembly Line in a JIT Production System," *Computers and Industrial Engineering*, 27 (1994), pp. 71-74. [C]

Yafie, Roberta C. "Profitability Becomes the New JIT Issue," *American Metal Market*, 105 (1997), pp. 28-30. [C]

Yan, H. "The Optimal Number of Kanbans in a Manufacturing System With General Machine Breakdowns and Stochastic Demands," *International Journal of Operations and Production Management*, 15 (1995), pp89-103. [N, K]

Yanagawa, Y., Miyazaki, S., and Ohta, H. "An Optimal Operation Planning for the Fixed Quantity Withdrawal Kanban System with Variable Lead Times," *International Journal of Production Economics*, 33 (1994), pp. 163-168. [K, N]

Yano, C. A. "Optimizing Transportation Contracts to Support Just-In-Time Deliveries: The Case of One Contracted Truck Per Shipment," *IIE Transactions*, 24 (1992), pp. 177-182. [P]

Yasin, M. M., Small, M., and Wafa, M. A. "An Empirical Investigation of JIT Effectiveness: An Organizational Perspective," *Omega*, 22 (1997), pp. 461-471. [E, H]

Yasin, M., and Wafa, M. A. "An Empirical Examination of Factors Influencing JIT Success," *International Journal of Operations and Production Management*, 16 (1996), pp. 19-26. [E]

Yavuz, I., and Satir, A. "A Kanban-Based Simulation Study of a Mixed Model Just-In-Time Manufacturing Line," *International Journal of Production Research*, 33 (1995), pp. 1027-1046. [K, S]

Yoo, S. "An Information System for Just-In-Time," *Long Range Planning*, 22 (1989), pp. 117-126. [C]

Youssef, M. A. "Measuring the Intensity Level of Just-In-Time and its Impact on Quality," *International Journal of Quality and Reliability Management*, 11 (1994), pp. 59-80. [Q, N]

Zangwill, W. "Eliminating Inventory in a Series Facility Production System," *Management Science*, 33 (1987), pp. 1150-1164. [C, N]

Zangwill, W. "From EOQ Towards ZI," *Management Science*, 33 (1987), pp. 1209-1223. [N]

Zangwill, W. "The Limits of Japanese Production Theory," *Interfaces*, 22 (1992), pp. 14-25. [C]

Zantinga, J. T. "Improvement in Spanish Factories: Towards a JIT Philosophy?," *International Journal of Operations and Production Management*, 13 (1993), pp. 40-48. [I]

Zapfel, G., and Missbauer, H. "New Concepts for Production Planning and Control," *European Journal of Operational Research*, 67 (1993), pp. 297-320. [C]

Zhao Xiaobo Z., and Zhaoying Zhou, Z. "A Semi-Open Decomposition Approach for an Open Queueing Network in a General Configuration with a Kanban Blocking Mechanism," *International Journal of Production Economics*, 60 (1999), pp. 375-381. [K, N]

Zhang, H., Kuo, T., and Lu, H. "Environmentally Conscious Design and Manufacturing: A State-of-the-art Survey," *Journal of Manufacturing Systems*, 16 (1997), pp. 352-371. [C]

Zhu, Z., Heady, R. B., and Lee, J. "A Simple Procedure for Solving Single Level Lot Sizing Problems," *Computers and Industrial Engineering*, 26 (1994), pp. 125-131. [C]

Zhu, Z., Meredith, P., and Makboonprasith, S. "Defining Critical Elements in JIT Implementation: A Survey," *Industrial Management and Data Systems*, 94 (1994), pp. 3-10. [C]

Zhuang, L. "Towards a More Economic Production and Just-In-Time Delivery System," *International Journal of Production Economics*, 36 (1994), pp. 307-313. [J, N]

Zhuqi, X., and Shusaku, H. "A Study on Sequencing Method for the Mixed-Model Assembly Line in Just-In-Time Production Systems," *Computers and Industrial Engineering*, 27 (1994), pp. 225-228. [C]

Zipken, P. "Does Manufacturing Need a JIT Revolution?," *Harvard Business Review*, 1991, pp. 40-50. [C]

Zolkos, Rodd "Property Loss: 'Just-In-Time' Approach Exacerbates Risk of Contingent Business Interruption," *Business Insurance*, 31 (1997), pp. 3-6. [C]

INDEX

About the Authors

MARC J. SCHNIEDERJANS is Professor of Management in the College of Business Administration, University of Nebraska-Lincoln. Author or coauthor of 10 books on various topics in management and more than 70 articles, he also has served as a consultant and trainer to a wide variety of business and government agencies, and as a technical expert for government organizations. His most recent Quorum Book, published in 1998, is *Operations Management in a Global Context*.

JOHN R. OLSON is an Assistant Professor at DePaul University in Chicago. Previously, he was a consultant and trainer for manufacturing and service organizations throughout the upper Midwest. He also has served as a project director, process engineer, and quality auditor for the state of Minnesota.

ISBN 1-56720-155-5

HARDCOVER BAR CODE